SLATE WASTE:

Engineering and Environmental Aspects

SLATE WASTE:

Engineering and Environmental Aspects

K. L. WATSON

B.Sc., Ph.D., C.Chem., M.R.I.C.

Senior Lecturer in Materials Science,
Department of Civil Engineering,
Portsmouth Polytechnic, UK

APPLIED SCIENCE PUBLISHERS LTD

LONDON

APPLIED SCIENCE PUBLISHERS LTD
RIPPLE ROAD, BARKING, ESSEX, ENGLAND

British Library Cataloguing in Publication Data

Watson, Keith Leslie
Slate waste.
1. Slate industry—Waste disposal
2. Slate
I. Title
620.1'32 TN958

ISBN 0-85334-880-4

WITH 26 TABLES AND 69 ILLUSTRATIONS

Printed in Great Britain by Galliard (Printers) Ltd, Great Yarmouth

Contents

Preface

This book comes at a time when the industrial community is consuming increasing quantities of raw materials at the same time as seeking to conserve its resources; this paradox can be partly resolved by more efficient recycling of industrial by-products and wastes.

The extraction of slate, and the manufacture of finished slate products, have left a considerable legacy of waste material. The history of research into slate waste utilization goes back at least a hundred years but even the current output of this material still outstrips its commercial use.

The literature contains many references to both actual and potential uses published during the course of the present century; and recent activities, viewing industrial by-products and wastes as potential resources, have led to a reappraisal of the slate waste problem over the last few years. This book gives an overall view of the current state of technical knowledge against the complex background in which the obvious advantages of making good use of this material often seem to be offset by more nebulous constraints.

K. L. WATSON

Acknowledgements

In presenting this book, I would like to acknowledge the many sources of information which I have used. These are, of course, noted as references at the end of each chapter and in the captions of tables and figures. However, there are a number of people and organizations to whom I would like to offer my special thanks.

I am particularly indebted to Mr R. H. Boyle, Director of Penrhyn Quarries Ltd, who has fostered my interest in slate waste utilization and who has provided me with a great deal of information. I would also like to express my gratitude to the following people for the information and other help which they have given me: Mr B. K. Bardhan-Roy (Jan Bobrowski and Partners); Dr R. N. Crockett (Mineral Intelligence Unit, Institute of Geological Sciences); Mr F. S. Dexter (Council for Small Industries in Rural Areas); Dr C. G. Down (Royal School of Mines, Imperial College of Science and Technology); Mr J. D. Evans (South West Water Authority); Mr N. H. Fox (Central Electricity Generating Board); Mr G. W. Hickson (Aerated Concrete Ltd); Mr K. H. McGhee (Commonwealth of Virginia Highway and Transportation Research Council); Dr P. J. Nixon (Building Research Establishment); Dr G. O. Phillips (The North E Wales Institute of Higher Education); Mr H. C. Robinson (Expanded Shale Clay and Slate Institute); Mr I. E. Rutherford (Llechwedd Slate Caverns); Dr J. C. Sheldon (Department of Physical Planning, Lothian Regional Council); Mr P. T. Sherwood (Transport and Road Research Laboratory); Mr A. Short (Comité Euro-International du Béton); Mr D. B. Snushall (Lemon and Blizard); Mr B. H. Spratt (Cement and Concrete Association); Dr R. N. Swamy (Department of Civil and Structural Engineering, University of Sheffield), Mr P. Titman (Mears Construction Ltd); and many of my colleagues at Portsmouth Polytechnic, particularly Mr J. Lowe and Mr N. Crighton of the Central Photographic Section and Mr I. Cross, Mr N. B. Eden, Mr P. Etherington and Mr E. J. Humby of the Civil Engineering Department.

I am most grateful to the following organizations who have allowed me permission to quote information from their publications cited in the text: The College of Earth and Mineral Sciences Experiment Station (The Pennsylvania State University); Delabole Slate (1977) Ltd; The Institution of Civil Engineers; The Organisation for Economic Co-operation and Development; The Transportation Research Board (Washington, D.C.); and The United States Bureau of Mines. The following have also most kindly allowed me to make use of their copyright material: The American Society for Testing and Materials, 1916 Race Street, Philadelphia, PA 19103 (refs. 111 and 112 in Chapter 3); The Building Research Establishment (Crown copyright material, indicated in the text, contributed by permission of the Director; reproduced by permission of the Controller, HMSO); The Cement and Concrete Association (Fig. 3.13); The Institute of Geological Sciences (Fig. 1.1); John Wiley and Sons, Inc. (references cited, from *Bituminous Materials: Asphalts, Tars, and Pitches* (Editor: A. J. Hoiberg), in Chapter 2).

There are also the many other important sources that I have cited, which space does not allow me to acknowledge separately here. I have included a number of references, that are only mentioned in passing, where I have noted corresponding abstracts for the benefit of the reader who may wish to consult these before deciding whether he needs to see the original sources. As will be apparent, I have found *Chemical Abstracts* to be particularly helpful for this purpose, and for identifying relevant publications; I can recommend this periodical as a useful aid to current literature searches related to various aspects of the subject.

Finally, I would like to thank my wife, not only for typing the final manuscript but also for enduring my constant preoccupation in preparing it.

1

The Slate Waste Problem

1.1. SLATE

The word 'slate' was originally derived from an old French verb *esclater* meaning 'to split' or 'to shiver to pieces', and it has been applied generally to any rock that can be cleaved or split into thin sheets.[1] More precise, scientific definitions refer to the geological origins of slate from argillaceous sedimentary deposits, or sometimes from volcanic ashes, that have been metamorphosed to give a rock that exhibits characteristic 'slaty' cleavage. However, the word is still used loosely, even in the stone and building industries, and it is also found to refer to the shaly material that is associated with coal.[2]

In this book, the word slate is employed in its more scientific sense and, with a few exceptions of relevant interest, generally refers to the material which is well known for its use in the form of roofing slates. This same material is, of course, found in many other building applications such as damp courses, walling, cladding, flooring and paving. It is, or has been, used for the manufacture of various products such as electrical switchboards, tombstones, billiard table tops, bench tops, blackboards and the old fashioned writing slates. Slate is also processed (by crushing and grinding methods) into granules and powder, which find many applications in a variety of different industries; slate granules are used for surfacing roofing felts and the powder is used as an inert filler in bituminous compositions, plastics and rubber, paints and other materials. Crushed slate has also been used as an aggregate for road construction and in concrete. Many slates 'bloat' at high temperatures, to form a cellular structure, and this expanded slate has been used successfully as a lightweight aggregate.

1.2. ORIGINS AND COMPOSITION OF SLATE

It is generally recognized that the majority of commercial slates are derived from fine material, formed by the weathering of rocks, that had originally been carried to the sea or into lakes by the drainage of surface water. Eventually the sedimentary deposits of this material became buried and were then subjected to movement and to heat and pressure beneath the surface of the earth; and the consequent 'metamorphism' resulted in the formation of slate. Clays and shales are also derived from sedimentary deposits and, although their subsequent geological history differs from that of slate, the gross chemical composition of all three types of material is usually similar—but metamorphism is the agency that is responsible for the mineralogical characteristics of slate. Although most common slates are derived from argillaceous sediments, there are some which have been formed from accumulations of fine-grained volcanic ash and dust.

Table 1.1. The gross chemical composition of most slates (from ref. 2).

	% by weight
SiO_2	45–65
Al_2O_3	11–25
FeO	0·5–7
Na_2O	1–4
K_2O	1–6
MgO	2–7
TiO_2	1–2

Table 1.1 indicates the range of chemical compositions of most slates, as given by Crockett.[2] North[1] referred to the similarity in the chemical composition of slate and shale and he provided analyses, which represent the averages for 20 slates of various geological ages and 51 shales of Palaeozoic age, which show remarkable correspondence.

However, chemical analysis alone gives an oversimplified picture. For instance, clays, shales and slates of similar chemical constitution have very different properties.[1] Dry clay can be crumbled into irregular pieces, and shale can be split along planes related to the bedding of the rock. However, commercial slate can be split into strong sheets along planes normal to the direction of maximum pressure to which it was exposed during its formation; these planes are not the same as the bedding planes, although they can occasionally coincide by chance. So, for practical purposes, it is necessary to consider the minerals present.

The most important mineral constituents of slate are, according to Crockett,[2] generally as shown in Table 1.2. Illite may also be present, together with minerals such as calcite, dolomite, pyrite, feldspar, graphite and other carbonaceous matter, as well as small quantities of other minerals.[2,3] Although a high proportion of the silica may be present in the

Table 1.2. The most important mineral constituents of slate (from ref. 2).

	% by weight
Sericite mica	38–40
Quartz	31–45
Chlorite	6–18
Haematite	3–6
Rutile	1–1·5

form of quartz, some of it is combined in the micaceous minerals illite, sericite and chlorite and in feldspar.[3] The minerals are distributed fairly uniformly and flakes of mica form a network structure enclosing the other constituents; these flakes are orientated parallel to one another, as a result of the pressure during metamorphism. Slate can therefore be cleaved parallel to the direction of this orientation; the cleavage properties, hence the commercial value of the slate for roofing purposes, depends on the uniformity of the orientation and the presence of other minerals within the microstructure which may interfere with the splitting process. The characteristics of different slates depend on the nature of the original sediments and the modification of their mineralogical constituents, as well as the formation of new ones, during metamorphism.

1.3. PROPERTIES OF SLATE

Perhaps one of the most significant properties of slate, as far as its commercial development was concerned, is its cleavage. This enables skilled craftsmen using simple tools (basically a mallet and chisel) to produce thin roofing slates by splitting blocks into thin sheets. The relative simplicity of the process, and the inherent properties of slate as a roofing material, led to the development of a major industry long before more sophisticated technology could offer the alternatives which were ultimately to contribute to the decline of the slate industry.

The way in which slate has been formed generally gives it low porosity, and the principal minerals that it contains are relatively stable and inert; this means that many slates are highly resistant to weathering. However, some examples can contain significant quantities of calcite or pyrite which are both subject to attack, particularly in industrial areas where rainwater may be mildly acidic. There are sometimes other impurities and structural features, such as relatively high porosity, which can also lead to the poor weathering properties of inferior slates.

Slate is obtained in a variety of shades of black, grey, blue, green, and sometimes purple and red; this range of sometimes very attractive colours has, of course, significantly contributed to the popularity of slate roofing in the past. Variegated colour patterns can be found and distorted fossils occasionally occur. The differences in colour reflect the mineralogical composition; for instance, ferric iron gives a red colouration, ferrous iron and chlorite give green, and finely divided pyrite as well as graphite can give dark grey or black.[4] Furthermore, the colour of some slate changes after a period of exposure to the weather.

Other properties of slate seem to have received relatively little attention in the literature although standards exist in a number of countries which specify requirements for some of its applications. As might be expected, the properties of different slates vary rather widely depending on their source; furthermore, differences in testing procedures can lead to variations in measured values. 'Typical' values are therefore virtually meaningless and, whenever a new application for a particular material is contemplated, it is essential to thoroughly evaluate its performance by appropriate methods. Nevertheless, bearing in mind the comments above, it is perhaps worth noting some of the information that has appeared in the literature.

One report,[3] in 1947, reviewed some of the published data that was available at that time and more has been published since. An average value for the specific gravity has been given as 2·86, compared with 2·57 for clay[1] (and, of course, the higher density can be attributed to the compression during metamorphism). The thermal conductivity of one particular slate was found to be similar to that of bauxite bricks and terra cotta, and somewhat lower than that of fire brick.[5] Provided that it contains no conducting veins of minerals such as pyrite and magnetite, the electrical properties of slate generally make it a satisfactory low tension insulator, particularly where its non-combustibility is required[6] (and, as noted in Section 1.1, it has been used for the manufacture of electrical panels). Mechanical properties vary greatly and some quoted values are given in Table 1.3.

Table 1.3. Some values, quoted in the literature, for the mechanical properties of slates.

Source	*Tensile strength* (MN/m^2)	*Modulus of rupture* (MN/m^2)	*Compressive strength* (MN/m^2)	*Shear strength* (MN/m^2)	*Modulus of elasticity* (GN/m^2)
Ref. 3	25	81	69–98		93
		54–69	71		60–90
		49–78	64		97
		54–74[a]			55–65[a]
Ref. 5	58	113	217		
Ref. 6	21–28		76–103	28	
Ref. 16		43			

[a] Depending on the dimensions of the test specimen.

Finally, apart from the properties already mentioned, many slates can be expanded by high temperature processing (as noted earlier), and molten slate can be 'blown' to produce mineral wool; these aspects will be discussed in more detail in later chapters.

1.4. THE SLATE INDUSTRY

The history of the use of slate in the construction of buildings is known to go back nearly two thousand years. The Romans used it in building their fort of Segontium, near Caernarvon in North Wales;[7] they clearly appreciated its versatility and used it for roofing and flooring and even as a lining material for the channels of a hypocaust—and it is tempting to suppose that the broken slate, used with other materials for metalling courtyards and the praetorium at Segontium, is evidence that the history of slate waste utilization dates back as far as the known use of slate as a building material.

Slate quarrying has been practised for many hundreds of years in Wales. King Edward I stayed in a house, at Nantlle, said to have been roofed with slates from the nearby Cilgwyn Quarry which is believed to date from the twelfth century.[7] It is believed that both Conway and Caernarvon castles, built in 1282–4, were originally roofed with slates[8] and, in another thirteenth century castle at Cardigan, slate was used as a walling material.[9] The slate quarrying industry has early origins elsewhere;[7] evidence, in the form of a municipal charter granted to quarrymen, suggests that it had begun in the Ardennes by at least the middle of the sixteenth century. In

England, the 'Old Delabole' Quarry in Cornwall has been worked since about 400 years ago[10]—and long before that, in 1187, 80 000 slates were shipped to Winchester from Devon.[2]

Increased urbanization and the concomitant requirement for buildings provided a stimulus for the slate industry which responded with a dramatic growth in roofing slate production during the nineteenth century. The quarries in Wales have probably been the most productive in the world[11] and the development of the Welsh slate industry has been particularly well researched.[7] The best Welsh slates are reputed to be among the finest available and, although individual quarrying enterprises in Wales tended to be small up until the middle of the eighteenth century, subsequent developments and rationalization led to the creation of a major industry which exported its products as far as Australia.

In 1898, the annual production of slate in the United Kingdom reached its peak, at 650 000 tonnes;[2] and nearly 80 % of this was produced in Wales. But the output declined rapidly until, in 1918, only 111 000 tonnes were produced. There was some post-war recovery to between about 200 000 and 300 000 tonnes per annum in the 1920s and 1930s; but production fell again during the Second World War and, despite a slight improvement afterwards, there has been a downward trend since then to a figure of around 60 000 tonnes per annum in the early 1970s. Analysis of recent production figures has shown that the output of roofing slate fell from 39 000 tonnes in 1967 to 17 000 tonnes in 1973; the output of other slate products (much of which is believed to be granules and powder) remained between 41 000 and 47 000 tonnes during the same period, although these figures may not be reliable.[2] (Expanded slate was recently produced in Wales, but the plant has been closed because of a pollution problem.[12])

Similar general trends have been noted in other parts of the world. In Finland,[13] for instance, production has declined and is nowadays mostly confined to the manufacture of slabs for paving and decorative purposes, but these are generally made from micaschist (i.e. slate that has been subjected to extreme pressure resulting in a crystalline rock, mainly of mica and quartz, with poor cleavage[1]).

As in the United Kingdom, the United States slate industry has declined during the course of the present century.[11] Between 1910 and 1958 the annual sales volume of roofing slate fell from 1 260 261 squares to 86 000 squares (a square being sufficient material to cover 100 ft^2 (9·29 m^2) of roof using a 3 in (76 mm) head lap). In 1910, the annual sales volume of mill-stock (such as blackboards, electrical panels and building products other than roofing slates) was 481 377 m^2 and reached 1 029 167 m^2 in 1925;

subsequently it fell, reaching 195 819 m^2 in 1945, but then recovered to 344 484 m^2 by 1958. The annual sales of other slate products, mainly granules and powder, were valued at $392 997 in 1910 and, after a fluctuation in the 1920s and 1930s, reached $8 818 209 in 1950; their value in 1958 was $6 326 000 but this figure includes flagstones and other building products. (Of course, trends expressed in terms of the value of sales can be misleading. For instance, the actual quantity of roofing slate sold in the

Table 1.4. Classification of types of slate products and quantities shipped or used by producers in the United States in 1972 and 1973 (from ref. 14).

	Thousands of tonnes	
	1972	*1973*
Roofing (including a small amount of house stone veneer)	11	11
Mill-stock, flooring and flagging, etc.	63	69
Expanded slate	1 152	991
Others (including powder and granules)	221	340

United States declined by a factor of 14·7 between 1910 and 1958, and the corresponding factor calculated in terms of the dollar value of sales was only 2·4; between the same years, the quantity of mill-stock declined by a factor of 1·4, but the value of the sales actually increased by a factor of 3·1.)

For comparison with the United Kingdom statistics quoted earlier, the quantities of slate shipped or used by producers in the United States in 1972 and 1973[14] are classified by the type of product in Table 1.4. The commercial significance of expanded slate is particularly evident.

So there is much evidence that the slate industry, in a number of countries at least, has undergone considerable change; and similar trends have been observed in other countries. There has been a decline in the production of roofing slates because of the competition offered by alternative materials such as mass-produced tiles (particularly in the United Kingdom) as well as roofing products based on bituminous compositions, asbestos/cement sheeting and corrugated metal; the increasingly high cost of the craftsmanship required to split the slates and the difficulty in maintaining a skilled labour force have also made an important contribution. However, slate is still used for roofing a few new buildings, and there is also a small replacement market. The decline has been partly offset by the manufacture of products such as mill-stock, flooring, flagging, granules and powder; some quarries have remained profitable, partly by the commercial development of these products and partly because of the closure of less fortunate competitors. The production of expanded aggregate is proving to be an important use for slate in the United States and, as will be seen in Chapter 3, in other countries too.

The examples quoted above should not be taken to imply that slate production is confined to the countries already mentioned; it is, or has been, produced in many countries on the continent of Europe and in Australia, Canada, India, Japan, New Zealand and South Africa as well as others.

1.5. THE EXTRACTION AND PROCESSING OF SLATE

The method of extracting slate depends very largely on the nature of the geological formation in which it is found. If the strata are steeply inclined, then the slate bed can often be followed by deep, open quarries where the ground is relatively level, or by terraces where the bed outcrops on a slope such as a mountainside.[1] If the strata are less steeply inclined, say at below 45°, then as the quarry is extended, progressively more overburden has to be removed and mining techniques become more profitable—but, in this case, good material may have to be left as pillars to support the roof. Even where the slate bed is very steeply inclined, vertical or near-vertical cleavage can make the quarry wall weak and unsafe; in one such instance,[11] deep shafts were sunk and the slate was removed by overhead stoping from lateral adits. However, in most cases, slate is extracted by means of open quarries.

The use of explosives is generally necessary to obtain blocks of slate for processing; different techniques are used and a great deal of experience has been accumulated in reducing fragmentation which can otherwise render large quantities of good material useless. A number of methods have been investigated to minimize wastage at the quarrying stage. For example, the use of the wire saw has proved to be successful in a number of cases; the moving wire, in the form of a continuous loop, is fed with a slurry of sand and water which acts both as an abrasive medium and as a coolant. The wire saw was introduced to the Pennsylvania slate industry in 1926 by the United States Bureau of Mines, and new developments in this technique were discussed in 1955.[15] Wire sawing was also introduced into a British quarry in 1948–9.[16] The use of rotary diamond saws and chain saws has also been investigated.[15]

For the production of roofing slates, the material from the quarry is first reduced to blocks of convenient size; usually these are then split by hand, along their cleavage direction using mallets and chisels, to form separate slates of the correct thickness. After splitting, the edges are trimmed to the required dimensions, either by hand or by machine, although sometimes the slates are split from blocks that have previously been sawn to the right size. Slate splitting machines have been developed[15] and they have sometimes been used with considerable commercial success, particularly in France, but many slates are unsuitable for mechanized splitting.

As mentioned earlier, a variety of other dimensioned products are manufactured. The modern slate mill is equipped with various machines such as wire saws, frame saws, circular saws (sometimes over 2 m in diameter), planing machines, polishers and buffers, depending on the range of its products; and slabs of slate are produced in a wide range of shapes and sizes with riven, sawn or polished surfaces according to their end use.

Some quarries produce granules and powder and are equipped with crushing, grinding and screening facilities and sometimes with air separating plant to control the fineness of the powder.

The production of expanded slate involves high temperature processing plant, often in the form of rotary kilns, as well as crushing and screening facilities. This will be discussed in more detail in Chapter 3.

1.6. SLATE WASTE PRODUCTION AND STOCKPILING

A very high proportion of the slate extracted by quarrying and mining has to be discarded as waste. In one case[17] it has been said that, in order to produce 1 tonne of roughly finished slates, about 12·5 tonnes of block slate is required (i.e. material prepared at the quarry face) and, to produce this quantity of block slate, as much as 50 tonnes of waste may have to be quarried. A recent survey in the United Kingdom[18] showed that the overall ratio of waste production to finished products is about 20 to 1, by weight, in England and Wales.

As North pointed out,[1] much of this waste cannot be avoided. For instance, weathered rock near the surface does not usually yield blocks of useful size. Furthermore, to secure good slate, it may be necessary to quarry associated material with poor cleavage. Slate with closely set joints (i.e. planes of weakness or actual cracks), and fragments from blasting are seldom of any use for making dimensioned products. Splitting and trimming yield considerable amounts of waste, generally in the form of small pieces, and useless offcuts are produced from sawing operations.

Figure 1.1 shows an aerial view of the Penrhyn Quarry in North Wales,

Fig. 1.1. The Penrhyn slate quarry. (Photograph by Hunting Surveys Ltd, reproduced by courtesy of the Institute of Geological Sciences.)

which is the largest surface excavation in Britain;[19] the photograph illustrates the enormous quantities of waste that have been tipped around the workings. The quarry, in the form of a segment of a circle, is about 2 km long;[17] the actual excavation covers an area of nearly 40 hectares and the overall area, including the workings and associated buildings as well as the waste tips, is over 200 hectares. Figure 1.2 shows a slate waste tip at Blaenau Ffestiniog photographed from ground level. And Fig. 1.3 illustrates individual pieces of slate waste; these can range from blocks weighing several tonnes down to small fragments.

The current production of slate waste generally tends to be much less than in former years; this is partly due to improved quarrying and manufacturing techniques but mostly due to the reduced volume of slate processed in many areas. Nevertheless, the legacy of waste in old tips

Fig. 1.2. Slate waste tipped at Blaenau Ffestiniog. (Crown copyright photograph reproduced by courtesy of the Director of the Building Research Establishment.)

Fig. 1.3. Pieces of slate waste as tipped. (Crown copyright photograph reproduced by courtesy of the Director of the Building Research Establishment.)

throughout the world is enormous. For instance, in England and Wales, the quantity of stockpiled waste from the slate industry is second only to that produced by coal mining;[18,20] there are between 300 and 500 million tonnes of slate waste in Wales, 15–20 million tonnes in the Lake District of England and about 15 million tonnes in Devon and Cornwall—and a total of over a million tonnes per annum are currently produced. There are probably about 70 million tonnes, and perhaps more, left over from the now defunct industry in Scotland.[21] Huge stockpiles also exist in the United States,[22] and it is estimated that over 70 000 tonnes are generated annually, although this rate is declining by about 10% per annum. In Norway the slate waste problem is very large[23] and there are considerable quantities in Finland[13] and in other parts of the world.

The disposal of waste from quarries, mines and their associated processing facilities has always been a major problem for the slate industry; the problem is, of course, intensified by the very large proportion of waste that is generated. From the financial point of view, the cost of waste disposal forms an important part of the overall production costs. At a more practical level, it becomes more difficult to find space for convenient disposal as increasing quantities of material accumulate around the quarry or mine although, in the case of mining, the waste can often be left underground. In some instances, the problem has been amplified where waste produced in the past has been tipped in areas which have subsequently needed reclearing for further quarrying operations. In one extreme case,[24] quarrying operations were abandoned because waste piled at the edge of the quarry threatened the safety of the men working below. Sometimes old workings have been backfilled with waste but this is not very often convenient or economic.

Many slate quarries are located in thinly populated, remote areas of poor agricultural value and there are sometimes few obvious incentives to clear the tips. But these same areas are often of outstanding scenic beauty with considerable value for tourism; the dismal appearance of the tips and workings in North Wales led to the north-west boundary of the Snowdonia National Park being drawn to exclude the quarrying areas of Bethesda, Llanberis and Nantlle—and the Blaenau Ffestiniog area was excluded from the park in such a way as to form an enclave within it.[25] In the Lake District in England, one quarry was closed because of its waste disposal problem.[18] And, of course, where tips occur close to centres of population, they are a permanent nuisance to the local inhabitants. So, as far as the community at large is concerned, the main objection to slate waste is aesthetic. From this point of view it is very difficult to quantify, in monetary terms, the benefit that would be obtained by removing, reducing or at least improving the appearance of the slate waste tips—and this point will be discussed later on.

However, the question of slate waste utilization is much broader than this and involves other issues; and, as will emerge during the course of discussion in this book, perhaps one of the most fundamental of these is the problem of diminishing resources.

1.7. THE UTILIZATION OF SLATE WASTE

Apart from the use of broken slate for metalling by the Romans, mentioned in Section 1.4, there is evidence that waste material has been utilized over the course of many years. The typical Snowdonia slate fence, using rough slabs set in the ground and wired together at the tops, probably dates back to the time of the Enclosure Acts;[26] and an early example of utilization for bulk fill was the construction of a railway embankment from slate at the Cilgwyn Quarry in 1895–7 to reach a new waste tip. The manufacture of slate slabs in Wales developed in the 1860s and provided scope for using material that lacked good cleavage properties. However, slab production was generally undertaken separately from the quarrying enterprises.[7]

At about the beginning of the present century, there were signs of systematic attempts to put slate waste to good use.[7] In 1895, the manager of the Penrhyn Quarry wrote that Lord Penrhyn had lost so much money in experiments with slate dust that he would be unlikely to support any more —and he also referred to an apparently worthless scheme for the manufacture of Portland cement. In 1919, the North Wales Development Company was set up to manufacture slate dust products, such as paper and tyres, but was wound up in 1921.

Work was carried out in the United States with a view to using slate waste in manufacturing cement and in producing slate granules for making roofing materials. In 1921, two United States Bureau of Mines reports[27,28] described investigations into the utilization of waste slate as a powdered filler in a variety of applications. Despite some initial setbacks, the commercial development of slate granules and powder proved to be successful and, as will be seen in the next chapter, these materials still find many applications in various industries.

In the early 1930s, development studies were carried out at the Building Research Station, in the United Kingdom, which showed that the heat treatment of slate can give an expanded lightweight material that is an excellent aggregate for making concrete.[29] As noted above, this application has aroused considerable interest in a number of countries and will be considered in detail in Chapter 3, together with the use of unexpanded slate as a dense aggregate.

There were other studies during the first part of this century and these perhaps culminated with a major investigation in the United States. In 1932, the legislature of Pennsylvania recognized the decline of the slate industry in that area and, in the same year, the Mineral Industries Experiment Station of the Pennsylvania State College was authorized to begin studies into slate. The programme was curtailed, because of lack of funds, until 1943 when the Pennsylvania General Assembly appropriated funds for research aimed at improving the state of the slate industry there. Fundamental studies, involving X-ray diffraction, differential thermal analysis and electron microscopy, provided new information about the nature of slate;[3] and this work was extended to the development of new products from slate waste including slate/lime brick, lightweight aggregate, resin-bonded moulded products, mineral wool, an acoustical tiling material and other materials. Subsequent studies involved a thorough investigation of operating practices in the slate industry[30] and of technical developments.[15] The report on technical developments discussed a range of different aspects, some of which were capable of reducing the amount of waste produced; these were concerned with sawing techniques (including chain, wire and rotary mill sawing as well as the use of the rotary diamond saw in the quarry), the production of masonry units from slate, mechanical slate splitting and underground slate mining techniques.

Many of the actual and potential applications for slate waste had been proposed by the late 1940s although some new ideas and extensions of old ones have appeared since that time. With the 1970s came a period of re-evaluation of the slate waste problem in a much broader context. The industrial community had become aware that it was consuming increasing quantities of raw materials at the same time as seeking to conserve its resources; and there was increasing recognition of the fact that this paradox could be partly resolved by more efficient recycling of industrial by-products and waste materials.

The construction industry consumes very large quantities of raw materials and it is perhaps in this area that slate waste could be used most effectively.

It is now becoming recognized that, in some countries, there are localized shortages of conventional road building materials, particularly in urban areas; these shortages are critical in some cases and it is anticipated that the problem will both intensify and become more widespread. Serious attention has therefore been paid to the use of slate waste in road construction (with utilization having been achieved in some areas), and this is discussed in Chapter 4.

The energy crisis has highlighted the better thermal insulation properties of lightweight concrete when compared with normal dense concrete; also, the lighter structures and foundations that can be used introduce other advantages that are relevant to conserving energy and materials—and, of course, the wider use of lightweight aggregate would provide an opportunity to develop the market for expanded slate. But, as will be seen in Chapter 5, there are a variety of other building products, based on autoclaved (i.e. high-pressure steam cured) calcium hydrosilicate materials, that could probably be manufactured from slate waste. Apart from dense products, such as slate/lime bricks, these include aerated lightweight materials with good thermal insulation properties.

There are a number of other applications for slate waste, both actual and potential, that do not readily fall into the categories already mentioned. Some of these are discussed in Chapter 6 and range from *ad hoc* uses, such as the production of blocks for sea defence work and for walling stone, to more tenuous prospects such as the potential of slate as a source of alumina for the manufacture of aluminium; and, again, there is the possibility of manufacturing an energy saving material, namely mineral wool.

Although there are various ways in which slate waste can be utilized there are many tips which, at the present time, could not be used economically. In such cases, reclamation is an alternative possibility that is discussed in Chapter 7.

Despite the considerable efforts that have been made to put slate waste to good use, the vast quantities throughout the world generally remain unchanged; the United Kingdom uses only a very small proportion of its current output and the stockpiles are slowly increasing—and this is happening in other countries too. The final chapter therefore considers some of the broader issues involved in promoting increased utilization; inhibiting factors and incentives are discussed in general terms.

1.8. HEALTH HAZARD

Before moving on to the next chapter it is particularly important to note that there is a health hazard associated with slate dust; it is a well-known fact that silicosis occurs among slate workers[31,32] and there are examples of workers in other industries which use slate powder who have developed this disease. The dust is generated, to a varying degree, in processing slate for any purpose and the risk often exists when slate-based materials are used for manufacturing other products—in any application, the hazard must always be properly assessed and the appropriate precautions must always be taken to protect health and safety.

REFERENCES

1. F. J. North, *The Slates of Wales*, 3rd Edition, The National Museum of Wales (1946).
2. R. N. Crockett, *Slate*, Mineral Dossier No. 12, Mineral Resources Consultative Committee, HMSO, London (1975).
3. *Properties and New Uses of Pennsylvania Slate*, Mineral Industries Experiment Station Bulletin No. 47, Pennsylvania State College (1947).
4. W. R. Power, 'Construction materials—dimension and cut stone', *Industrial Minerals and Rocks* (*Nonmetallics other than Fuels*), 4th Edition (Editor-in-Chief: S. J. Lefond), American Institute of Mining, Metallurgical, and Petroleum Engineers, Inc., New York (1975), 157–74.
5. R. H. O'Neill, 'The geological pedigree of Welsh slate—its many advantages over other roofing materials', *The Quarry Managers' Journal*, **43**(9) (Sep. 1959), 345–7.
6. A. E. Williams, 'Slate', *The Mining Magazine*, **89**(6) (Dec. 1953), 342–4.
7. J. Lindsay, *A History of the North Wales Slate Industry*, David and Charles, Newton Abbot (1974).
8. I. Griffith, 'Recent developments in the slate industry of North Wales', *The Quarry Managers' Journal*, **16**(3) (Jun. 1933), 87–92, 98.
9. S. Herbert, 'A temporary solution to a monumental problem', *Industrial Diamond Review*, **35** (Dec. 1975), 431–3.
10. J. Setchell, 'The Old Delabole Slate Quarries, Cornwall', *The Quarry Managers' Journal*, **16**(3) (Jun. 1933), 93–7.
11. O. Bowles, 'Slate', *Industrial Minerals and Rocks* (*Nonmetallics other than Fuels*), 3rd Edition (Editor-in-Chief: J. L. Gillson), American Institute of Mining, Metallurgical, and Petroleum Engineers, Inc., New York (1960), 791–8.
12. J. C. M. Forrest, 'The emerging potential of lightweight aggregate concrete', *Concrete*, **9**(12) (Dec. 1975), 17–19.
13. M. Eerola (Road and Traffic Laboratory, Technical Research Centre of Finland), private communication.
14. H. J. Drake, 'Stone', *Minerals Yearbook 1973*, *Volume I* (*Metals, Minerals and Fuels*), Bureau of Mines, United States Department of the Interior (1975), 1153–71.
15. F. D. Hoyt, *What's New in the Pennsylvania Slate Industry*, Mineral Industries Experiment Station Bulletin No. 66, The Pennsylvania State University (1956).
16. 'The quarrying of Westmorland blue slate', *The Quarry Managers' Journal*, **48**(11) (Nov. 1964), 445–54.

17. Sir Alfred McAlpine, 'North Wales slate', *Building Materials* (Jun. 1971), 36–7.
18. W. Gutt, P. J. Nixon, M. A. Smith, W. H. Harrison and A. D. Russell, *A Survey of the Locations, Disposal and Prospective Uses of the Major Industrial By-products and Waste Materials*, Building Research Establishment Current Paper CP 19/74, Department of the Environment (Feb. 1974).
19. Sir Kingsley Dunham, 'Geological setting of the useful minerals in Britain', *Proceedings of the Royal Society of London, A*, **339**(1618) (Aug. 1974), 273–88.
20. E. H. Coleman and P. J. Nixon, *A Survey of Possible Sources in Wales of Raw Materials for the Manufacture of a Lightweight Expanded Slate Aggregate*, Building Research Establishment Current Paper CP 78/74, Department of the Environment (Aug. 1974).
21. E. M. Lawson and P. J. Nixon, *A Survey of the Locations, Disposal and Prospective Uses of the Major Industrial By-products and Waste Materials in Scotland*, Building Research Establishment Current Paper CP 50/78, Department of the Environment (Jun. 1978).
22. W. Clayton Ormsby (Chief, Chemistry and Coatings Group, Federal Highway Administration, United States Department of Transportation), private communication.
23. A. Grønhaug (Public Roads Administration, Oslo), private communication.
24. J. E. Conley, H. Wilson, T. A. Klinefelter *et al.*, *Production of Lightweight Concrete Aggregates from Clays, Shales, Slates, and Other Materials*, Bureau of Mines Report of Investigations No. 4401, United States Department of the Interior (Nov. 1948).
25. C. T. Crompton, 'The treatment of waste slate heaps', *Town Planning Review*, **37** (Jan. 1967), 291–304.
26. C. G. Down (Department of Mineral Resources Engineering, Royal School of Mines, London), private communication.
27. O. Bowles, *Slate Dust in Asphalt Road Surface Mixtures*, Bureau of Mines Report of Investigations No. 2230, United States Department of the Interior (Mar. 1921).
28. O. Bowles, *The Utilization of Waste Slate as a Filler*, Bureau of Mines Report of Investigations No. 2283, United States Department of the Interior (Sep. 1921).
29. E. H. Coleman, 'A lightweight aggregate produced from slate waste', *Concrete and Constructional Engineering*, **31**(1) (Jan. 1936), 47–51.
30. C. W. Stickler, W. F. Mullen and A. Wayne Bitner, *Industrial Studies of Pennsylvania Slate Production*, Mineral Industries Experiment Station Bulletin No. 58, Pennsylvania State College (1951).
31. W. K. C. Morgan and A. Seaton, *Occupational Lung Diseases*, W. B. Saunders Company, London (1975).
32. D. Hunter, *The Diseases of Occupations*, 5th Edition, The English Universities Press Limited, London (1975).

2

Slate Granules and Powder

2.1. INTRODUCTION

Granules and powder are long-established products of the slate industry. Although they can both be produced from waste material, the requirements of the market often dictate that the slate must be extracted specifically for their manufacture; for instance, granules in particular are mostly required in specified colours.

The markets for granules and powder tend to be quite separate. The generally inert nature of many slates has led to the widespread use of the powder as a filler and extender in a broad range of products. On the other hand, the major proportion of granule production is believed to be consumed by the roofing industry.

2.2. SLATE GRANULES

Slate granules have been used for surfacing bituminous roofing felt since about the beginning of the present century; by that time, the manufacture of granule-surfaced bituminous roll roofing had just begun and this was shortly followed by the production of shingles made by cutting the roll material to the required sizes.[1] The market expanded and in the United States, by the mid-1930s, about 50% of all the roofing materials used were based on bituminous compositions;[1] and an estimate in 1968 gave a figure of 80–85% of all roofing installed there.[2]

Figure 2.1, which illustrates a typical commercial product, shows that the granules have a platey form.

The coarser material obtained from screening slate powder was already being sold for use as roofing granules in 1906.[3] The early success and the subsequent commercial development of slate for this purpose stemmed from a number of factors. First, the granules provided attractive surface finishes that were available in a variety of natural slate colours; and these were the colours already associated with the traditional high quality slate roof. Furthermore, the slate grains were found to bond well to the bituminous material and their flat shape provided good cover to the surface; by careful attention to the sizing of the granules, maximum cover was achieved—and this, in turn, improved the fire resistance of the material.

Fig. 2.1. Slate granules. (The small divisions of the scale represent 1 mm.)

Also it became apparent that the opacity of the granules was important in protecting the underlying bitumen from the harmful effects of sunlight which otherwise caused the material to become brittle and to crack.[4]

As the market for granule-surfaced roofing materials developed, a variety of other natural and synthetic granules were used and the choice of colour became a more significant factor, particularly in the United States; various colouring processes were developed and by 1970 over 150 patents had been published.[4] By recent standards, natural slate colours tend to be regarded as dull and, even as far back as 1947,[3] the artificially coloured granules had captured a substantial part of the market from slate; furthermore, they possessed advantages which ensured that they retained it. Various methods of colouring slate granules have been discussed in the literature, as well as treatment with molten salts to modify their surfaces and to make them more suitable for colouring;[3] also, by coating slate granules, it has been possible to reduce the formation of dust by abrasion, associated with bulk transportation, that can be a problem.

The use of coated granules is now very firmly established in the United States but recent technological and market trends have tended to severely limit the minerals that can be used as base materials;[4] nevertheless, 'natural' uncoated granules, including those made from slate, still accounted for 22% of the total tonnage shipped in the United States in 1970.

One British manufacturer has developed a large market for natural slate granules, both at home and abroad;[5] the size and shape is regarded as being very important and this company has found that its customers prefer flat, disc-shaped granules of 8–24 mesh size.

It has been mentioned earlier that colour requirements often mean that slate must be quarried or mined specifically for granule production and, if waste material is used, then it must be carefully selected to meet the needs of the market; thus there appears to be little prospect, at present, of significantly reducing even the current production of waste from other slate products by the manufacture of roofing granules.

However, other commercial outlets seem possible. Like slate powder, which is discussed in the next section, granules have been used as a filler in artificial stone products based on polymeric binders, especially in applications such as the manufacture of resin-bonded cladding tiles; and they have also been used, to a limited extent, as a fine aggregate in certain cast concrete products. (Expanded slate granules of the kind used as lightweight aggregates in concrete, described in more detail in the next chapter, have been used in other constructional applications. Two recent Czechoslovakian patents have described the use of expanded slate, among other materials, as an aggregate for special anticondensation and diffusion mortars[6] and in insulating compositions.[7])

A further application for dense (i.e. unexpanded) slate granules, which may prove to be significant, is their use as a carrier medium for pesticides. Active pesticide compounds are generally applied in diluted form to ensure uniform distribution and to improve their effectiveness,[8] and the use of water-based sprays and of mineral-based dusts are both well known for this purpose; mineral granules had been introduced as a carrier medium by the early 1950s in the United States and in the early 1960s in the United Kingdom. Sumner[8] pointed out that granular carriers have many advantages over sprays and dusts; granule application can be more precise and is less susceptible to the effect of wind, the active chemicals tend to be released more slowly, safety aspects are improved and there are biological advantages too. Sumner explained that the initial acceptance of granular carriers was slow in Europe, mainly because of the lack of specialized machinery required but partly because of reservations about cost and also because of the limited range of pesticides initially available in this form. However, interest began to develop and, in the early 1970s, the quantities used began to increase.

The ranges of granule size used have varied from 60–90 mesh to 8–15 mesh but 30–60 mesh has found general acceptance;[8] a variety of minerals may be employed, depending on the type of granules to be produced—and hard, inert, non-absorbent materials such as slate can be used when the active chemical is applied to them as a coating. (A United States patent[9] has described the use of such granules, including those made from slate, as a carrier for liquid pesticides.)

The price of slate granules may prove to be advantageous in this application. Sumner quoted an average price of about £40 per tonne, in Western Europe, for a reasonably good granular carrier; at about the same time (1974), slate granules were being sold for £5–£6 per tonne in the United Kingdom.[10]

2.3. SLATE POWDER

2.3.1. The nature and general uses of slate powder

Technical publications relating to the slate industry indicate that there is a large number of general applications that have been suggested, at one time or another, for slate powder as an inert filler and extender; and these have included the manufacture of the wide range of products listed in Table 2.1. Some of these uses are now obsolete because of new technology or because of the current availability of more suitable materials; some are not viable because of the health risk associated with the handling of slate powder; and others seem to have been speculative or have failed to develop commercially for technical or economic reasons. Nevertheless, there have been, and there still are, significant markets and the technical literature reveals that considerable efforts have been made to develop some of these applications.

Apart from its mineralogical characteristics, the particle size distribution and particle shape of a mineral filler are often of some importance for particular applications. Slate powder is commercially available in a wide range of particle size distributions; for instance, one British company supplies grades varying from 50 % below 0·003 mm to 50 % below 0·03 mm. Figure 2.2 shows the particle size distribution of a commercial slate powder that was determined using different methods;[11] the small divergence

Table 2.1. Some of the products, mentioned in the literature, for which slate powder has been used or considered for use.

Abrasive soaps
Adhesives, putties and sealing compositions
Artificial leather
Bituminous mixtures of various types
Cleaning, polishing, sweeping and washing compounds
Pencils and crayons
Expansion jointing
Explosives
Fertilizers, insecticides, pesticides and disinfecting powders
Flooring compositions
Gramophone records
Linoleum and oil cloth
Oil and grease absorbents
Paints and surface coatings
Paper of various types and cardboard
Plastics products
Rubber and vulcanite products
Synthetic stone, slates, cladding and tiles
Thermal and electrical insulating materials
Tooth cleaning powders and various cosmetics
Wallboard
Window shades
Wood filler

between the Coulter counter and Andreason pipette methods at small particle sizes may possibly be due to the effect of particle shape on the results obtained by sedimentation methods.

The platey nature of the slate granules shown in Fig. 2.1 may often persist down to the small sizes which characterize powders and, as will be seen later, this can have an important effect in filler applications; furthermore, the particles in some powders tend to be elongated. Figure 2.3 shows four pairs of consecutive sieve fractions, above 0·063 mm, from two commercial materials; sample B generally appears to contain more elongated particles than sample A. And, at the finer end of the scale, Fig. 2.4 shows material below 0·063 mm taken from some commercial samples.

Many slates have low water absorption characteristics and this can be advantageous in a number of applications for fillers. Figure 2.5 shows the results of standing various mineral fillers over water in a sealed container;[12] the results are influenced by the fineness of the filler, which is not taken into account here, but the figure gives a general comparison and indicates the

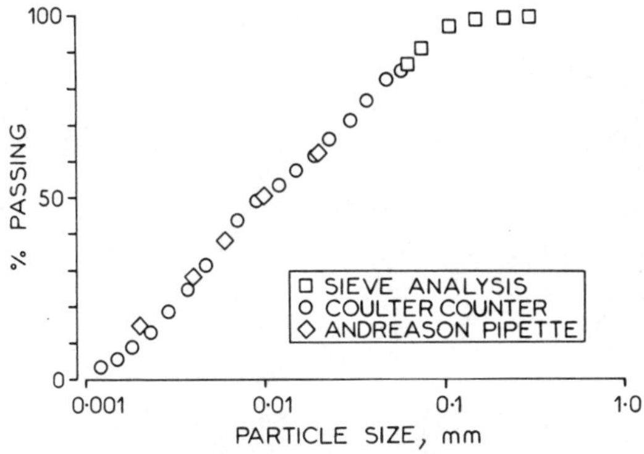

Fig. 2.2. The particle size distribution of a commercial slate powder (from ref. 11).

performance of a particular slate filler of which 99% passes a 0·053 mm sieve.

Most of the published technical investigations into the uses of slate powder seem to have been in the field of bituminous compositions, plastics and rubber compounds, paints, fertilizers, insecticides and pesticides, together with a relatively small number of other miscellaneous applications. These developments are discussed in the following section. There has also been some work aimed at investigating the feasibility of using slate powder as an admixture in concrete mixes and in autoclaved calcium silicate materials and these are also discussed in later chapters.

2.3.2. Slate powder in bituminous compositions

Bitumens are very complex mixtures of hydrocarbons and related compounds; they are brown or black in colour and, under normal conditions, take the form of solids or viscous liquids which progressively soften or become less viscous as their temperature is raised. They sometimes occur in natural deposits or, more generally, they may be manufactured as products from petroleum refining processes. Their usefulness stems mainly from their waterproofing and adhesive properties; and their ready availability and relative cheapness ensures their continued use in various building applications.

Naturally occurring bituminous materials were being used as long ago as 3000 B.C.; these 'natural' asphalts contained mineral constituents as contaminants which, in fact, behaved as fillers and they were employed as waterproofing agents and as mastics for brick-laying. When bituminous

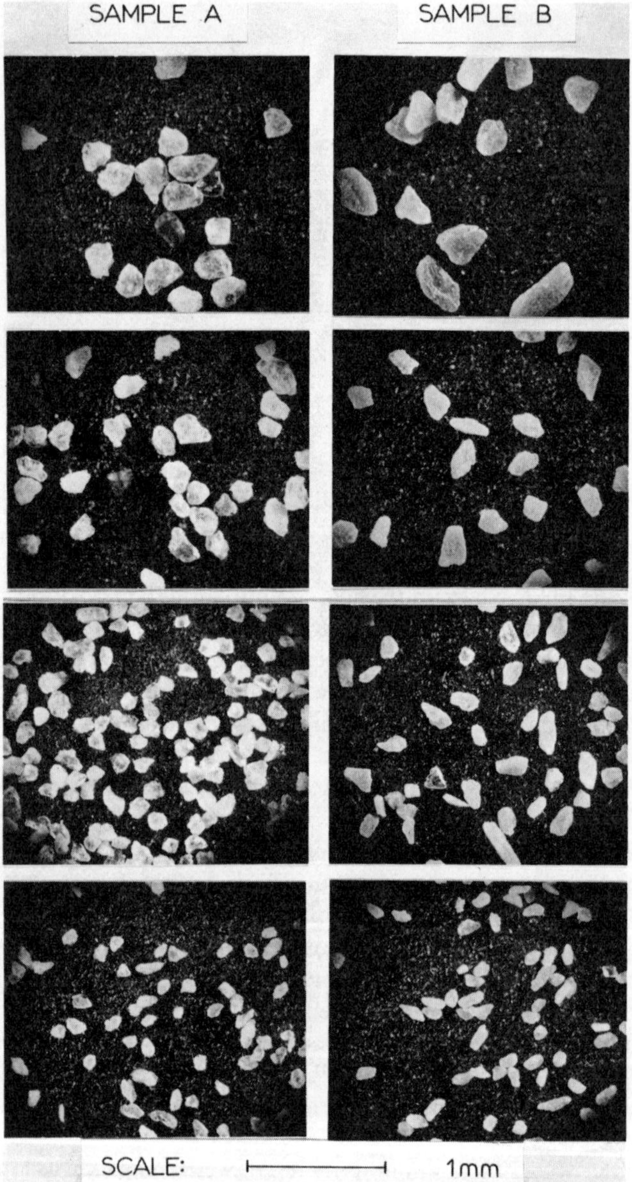

Fig. 2.3. A comparison of the shape of particles, within consecutive size ranges above 0·063 mm, taken from two commercial slate powders.

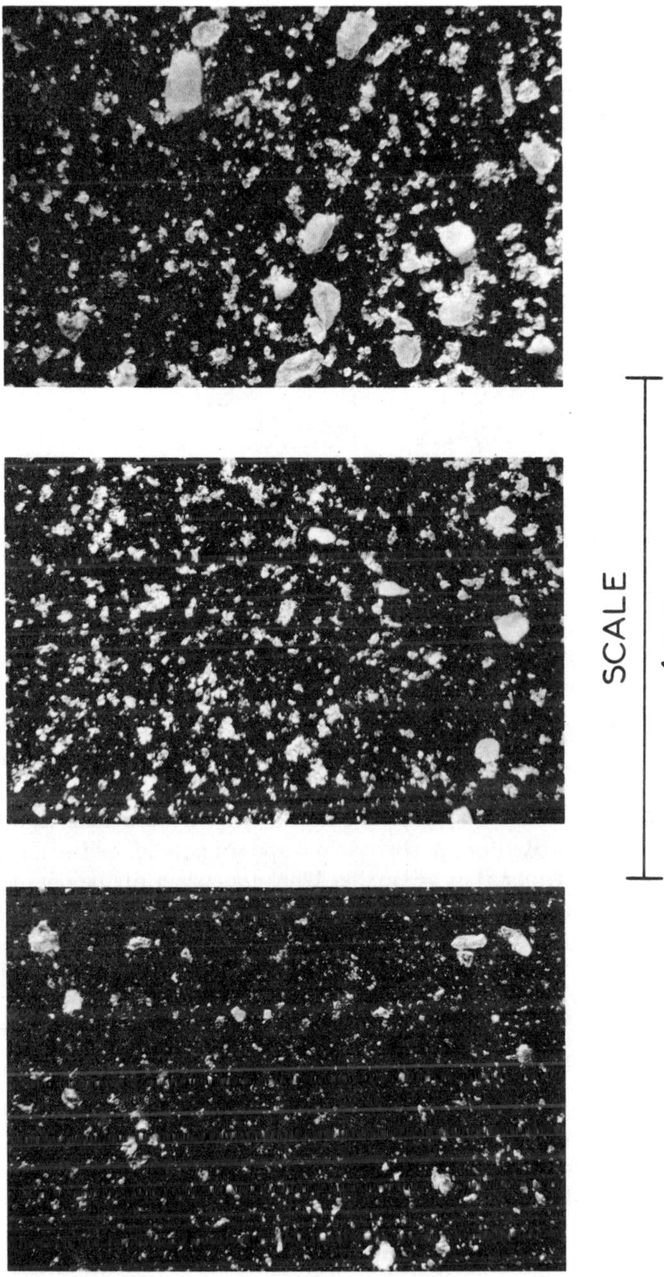

Fig. 2.4. A comparison of particles, passing a 0·063 mm sieve, taken from samples of commercial slate powders.

materials were first produced from petroleum refining processes, their properties were often found to be unsatisfactory; but it was discovered that their behaviour could be improved by compounding them with inert fillers —and soon the industry developed a better understanding of how their properties could be modified to suit a wide range of particular applications by means of the use of the appropriate type and quantity of filler.

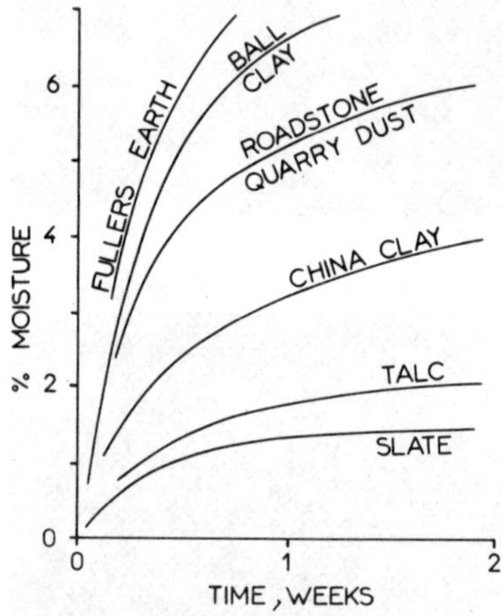

Fig. 2.5. The increase in moisture content, with time, for various fillers standing over water (from ref. 12, by courtesy of Delabole Slate (1977) Ltd).

Fillers are finely divided mineral materials. They are characterized by their particle size distribution, particle shape and, of course, by the nature of the mineral from which they are made. They may generally be regarded as having particle sizes ranging up to about 0·075 mm;[13] and it may be helpful to note that, in the context of bituminous materials, the fine aggregate (i.e. sand) may generally be regarded as consisting of particles ranging from about 0·075 mm up to 2 mm, and the coarse aggregate as particles from 2 mm upwards. The particle shape of many fillers is more or less isodimensional although some, including slate particles, may have a plate-like form and others, like asbestos, are fibrous.

One of the more important effects of fillers in bitumens is the modification that they produce in flow characteristics of the final mix. For instance, Fig. 2.6 illustrates the way in which viscosity increases with the content of short asbestos fibres and of a limestone filler. This example is derived from data, published by Shepherd,[14] which was obtained at 390 °F

(199 °C) using a rotating cylinder viscometer, and the maximum filler contents were about 7·5% and 70% for the asbestos and the limestone respectively. The figure illustrates that the viscosity increase is greatly influenced by particle shape, the fibrous filler being dramatically the most effective at low concentrations. Saal[13] has provided data showing that, at a volume concentration of 0·07, a spherical filler increases the viscosity of a bitumen by a factor of 1·25 compared with 1·30 for slate powder; the

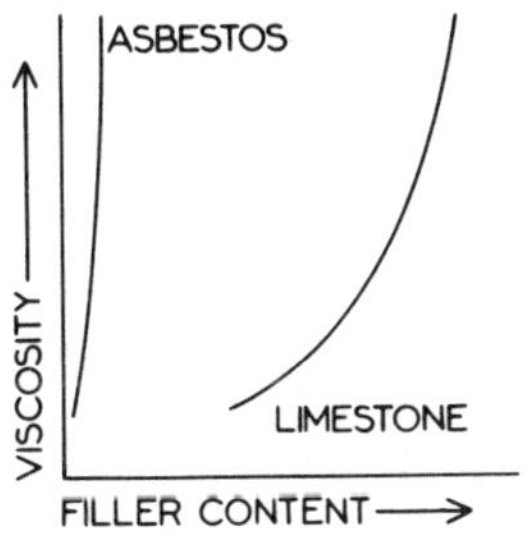

Fig. 2.6. The general effect of filler content on the viscosity of asbestos- and limestone-filled bitumen (after Shepherd[14]).

corresponding factors for micro-asbestos and short-fibred asbestos are 1·45 and about 30 respectively. This seems to suggest that slate powder, or at least the particular sample used to determine this factor, behaves similarly to an isodimensional filler in its effect upon viscosity. The relationship between filler content and the viscosity of the mixture is very complex; different mechanisms appear to operate at different levels of concentration.[13,14] The fineness of the filler is a factor which influences its stiffening effect (and the bulk density of filler suspended in benzene has been used as a simple measure of fineness in this context[15]).

Apart from affecting the viscosity and other related properties such as indentation resistance, the addition of appropriate types and quantities of filler can improve the strength and other mechanical properties of bitumens. Fillers may also be used to increase the resistance of bitumens to weathering; this effect is partly a result of the mechanical reinforcing effect which reduces cracking due to dimensional changes and, for opaque fillers such as slate powder, it is partly because of the shielding of the bitumen from the damaging effect of sunlight—also there are less well understood effects.[14] Fillers can be employed to increase the fire resistance of bitumen; this may be the result of straightforward effects such as dilution, reduced hot flow and reinforcement of the burnt residue—or of more subtle effects resulting from the use of intumescent fillers or materials that exhibit endothermic transitions.[14] Furthermore, the use of a cheap filler can reduce the unit cost of the finished product.

Bituminous materials are now used in very large quantities, particularly by the construction industry; major uses include road construction, roofing and a variety of waterproofing, coating and sealing applications. The consumption of fillers for these purposes is therefore considerable; for example, in the United States, it has been estimated that about 550 000 tonnes of filler material were used in 1970 for the manufacture of roll roofing and bituminous shingles alone.[16]

There is, therefore, considerable scope for using slate powder in bituminous compositions and a large amount of development work has been done but, before considering actual applications, it may be helpful to consider the effect of slate filler on the properties of bitumens.

There are two very important tests which, for many years, have been widely used to characterize bitumens; these are the penetration test, which amounts to an empirical measure of viscosity, and the softening point test which indicates the temperature at which the bitumen reaches a degree of 'softness' defined by the test procedure. Both of these tests have been used in investigations involving slate filler.

In the United States and in the United Kingdom, the essential measurement provided by the respective standard penetration tests is the distance, expressed in tenths of a millimetre, that a needle penetrates the sample of bitumen (maintained at a standard temperature) under specified conditions; the needle is loaded with a given weight and is allowed to move downwards, under gravity, into the sample for a prescribed period. This test is used to define various grades of bitumen and it is also employed for quality control purposes. Different time, temperature and loading conditions are sometimes used when extreme penetration values are obtained with the standard procedure.

The penetration test is recognized as being not very suitable for characterizing bitumen/filler systems since coarser grains can interfere with the penetration of the needle. Nevertheless, it can provide useful comparative data from such compositions if due care is taken—and the penetration value is usually found to decrease with increasing filler content in the general manner indicated by Fig. 2.7.

Kunath[17] used this test, among others, in an investigation into the properties of bituminous compositions containing 'activated' fillers with surfaces pretreated during the grinding process. These fillers (prepared from two different limestones, a granite and a phyllitic clay slate) were produced by grinding each mineral with bitumen and/or other organic materials (usually stearic acid) which covered the grain surfaces, including those newly formed during comminution. The slate was ground with 0·25 % stearic acid and 0·25 % of a road construction grade bitumen. For all the fillers, the proportion of grains larger than 0·09 mm was very low and the slate filler was the finest (the grain size distribution curve suggesting that over 50 % was below 0·02 mm). Both activated and untreated fillers were

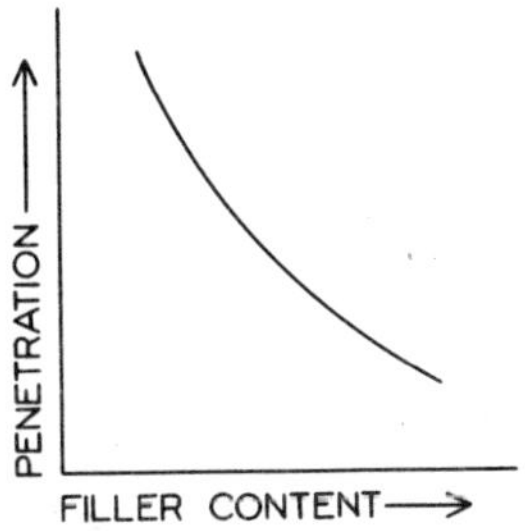

Fig. 2.7. The general effect of filler content on the penetration value of filled bitumen.

used to prepare various bitumen 'mortars' with filler/bitumen ratios ranging from 0·67 to 2·33.

The penetration measurement of each mix was expressed as the ratio obtained by dividing the penetration value of the mortar by the value of the bitumen alone; the values of this ratio were then plotted, for each filler, against the filler volume concentration. As might be expected, penetration decreased with increasing filler content in each case; but generally the mortars incorporating activated slate gave lower penetration values than the others, particularly at low volume concentrations.

Softening point tests can be devised to give a sharp 'end-point' measurement that provides a characteristic value for a particular bitumen; solid bitumens are mixtures of different molecules of various sizes and, of course, do not show a sharp transition corresponding to the melting of crystalline solids—instead, they progressively soften as their temperature is raised. The 'ring and ball' test is widely used to determine their softening points. The bitumen is cast into a metal ring to form a flat disc, then a ball bearing is placed centrally on its upper surface and it is suspended horizontally in a suitable liquid; the temperature of the liquid is then raised at a constant prescribed rate and, as the bitumen softens, the weight of the ball helps it to flow downwards through the ring. When the bitumen just touches the surface of a horizontal plate set at a fixed distance below the ring, the temperature of the liquid is recorded as the softening point. Like the measurement of penetration, this test is primarily intended for unfilled bitumens although it has been used to assess the effect of fillers; Fig. 2.8 indicates the general way in which the softening point is raised as the filler content of the bitumen is increased.

Shepherd[14] pointed out that, for conventional fillers, the softening point is raised almost linearly as the filler content increases to about 40–50 % by

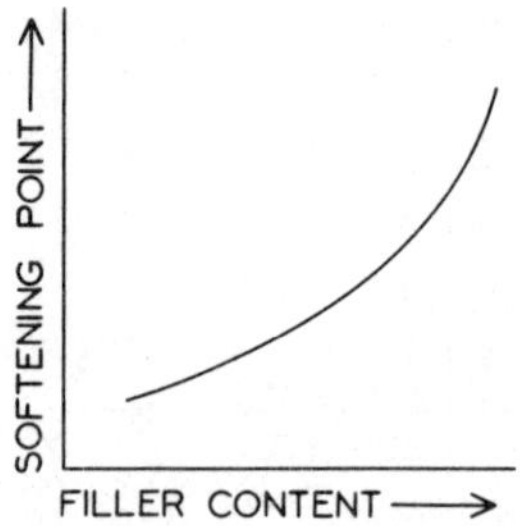

Fig. 2.8. The general effect of filler content on the softening point of filled bitumen.

weight; above this value the softening point is increased at a progressively higher rate. He also pointed out that highly effective fillers, such as asbestos, exhibit a short linear range up to about 5–20 % content. He presented graphical data showing a linear range of around 30 % for a diabase filler and for slate powder; but the slate provided a greater rate of increase in softening point with filler content both within the linear range and beyond it up to 60 % (although it should be noted that the bitumens used were not identical in the two cases). Furthermore, Shepherd quoted data showing that, up to a 40 % filler content, the increase in softening point of a coal tar due to a 1 % increase in slate powder was found to be somewhat less than 0·5 °F (0·28 °C), which was markedly greater than for limestone but considerably less than for asbestos fibre.

Kunath[17] also used the ring and ball test for the work described above in connection with the penetration values of filled bitumens. He provided graphical data showing the increase in softening point with volume concentration of the filler; for the types of filler he investigated, the form of the graphs indicate that the minimum volume concentrations used (i.e. down to about 0·30–0·36) were above the linear range. And the softening point versus concentration curves for both the activated and the untreated slate fell well above the corresponding curves for all the other fillers; the test showed very little difference between the two slate fillers but they both clearly increased the softening point to a greater extent than all the other materials at the same concentrations. Kunath also measured the viscosity of his materials over the temperature range of 20–90 °C; the slate fillers gave considerably higher viscosities than the others at a filler/bitumen ratio of 1·5, and generally less markedly higher viscosities at 1, but at 0·67 the differences were less obvious. (However, it should be noted, as mentioned above, that the slate filler was the finest of those investigated.)

Shepherd[14] provided evidence that several aspects of the practical performance of bitumens are improved by the addition of slate powder. For example, in one instance, the accelerated weathering life was increased from less than 80 cycles to failure (ASTM D 529-59 T, Cycle A) for the unfilled material to about 140 cycles for materials based on the same bitumen but containing 40 % and 60 % slate powder; from 0 % up to 40 % the improvement seems to have been more or less linear. He also quoted data, from Walther,[18] showing that the impact resistance of a coal tar was improved to a greater extent by a slate filler than by limestone (but much less than by asbestos); he also provided graphical data showing a linear increase in the impact resistance of a bitumen as the slate content was increased up to 60 %—and the improvement was greater than that provided by a diabase filler (although the bitumens used were different in each case). Fillers generally reduce the pliability of many bitumens although, in some cases, increased pliability can be achieved by the addition of only a small quantity of filler. Shepherd interpreted this increase in terms of higher shear strength and reduced brittleness; and he provided data showing such an improvement for a bitumen of relatively low pliability containing 20 % slate powder (although a further increase in filler content beyond this was shown to have an adverse effect).

Alexander and Tarver[19] discussed the permeability to moisture and the water absorption properties of bitumens. They pointed out that generally there is no close relationship between these properties and that permeability is the more important of the two; low permeability is more significant in determining the use of bitumens in most of their applications—water absorption, in itself, is of less importance and although other properties may be affected by it such effects are usually relatively minor. Alexander and Tarver found that the addition of 20 % slate powder to a bitumen led to a decrease in permeability (the permeability constant having been reduced from 10 to 7·9) and an increase in water absorption (from 3·7 to 4·7 % under the conditions that they employed). They also quoted evidence, from Slotboom,[20] showing that the addition of 30 % by weight of slate powder led to lower water absorption values in both tapwater and seawater than the addition of the same quantity of micro-asbestos, cement, limestone and gypsum; these materials gave increasingly higher absorption values, in that order (apart from the cement filler in seawater, which gave particularly high values), when measured at intervals over a period of 10 years—and tapwater generally gave considerably higher values than seawater. Shepherd[14] quoted data, from Walther,[21] showing that the water absorption (after 40 days' immersion) of a coal tar and a bitumen was unaffected by the addition of 40 % slate powder; but the same quantity of limestone gave a water absorption value that was increased by about 40 % in the case of the coal tar and by about 30 % for the bitumen. He also gave another example of a bitumen which showed a linear increase in water absorption with slate content; 60 % slate slightly more than doubled the water absorption value of the unfilled material after 26 weeks' immersion.

This discussion suggests that, although some general doubts have been expressed about the use of slate powder as a filler in bitumens,[17] it compares very favourably with other materials in a number of respects. The many publications which refer to applications in which it is used, particularly in the patent literature, suggest that it does have commercial value in this type of material.

In 1921, Bowles[22] published one of the earliest accounts of a systematic investigation into the use of slate powder as a filler in bituminous compositions; this work was aimed towards the utilization of slate waste in road surfacing materials. Bowles pointed out that, although powdered limestone and Portland cement were used as fillers in such mixtures, there was some opinion at that time that slate powder was also a good filler. To obtain data to support this, slate powder (of which 99·3 % passed through a 100 mesh screen and about 80 % through 200 mesh) was used to prepare specimens containing 20 % by weight of the bituminous binder; and similar mixtures were prepared using limestone and Portland cement powders. Impact tests gave very similar results for all three materials but the slate powder gave superior results in compression tests which provided a measure of the 'cementing value'. The same slate powder was used to prepare standard pavement mixtures containing 10 % by weight of bituminous binder, 13 % slate powder and 77 % graded sand. Compared with the limestone and Portland cement fillers, the slate powder provided marginally better impact values but the compression properties were slightly inferior to those using limestone but better than those from Portland cement. The slate powder was found to contain about 15–25 % more fine material, constituting effective filler, than the other materials and its cost was considerably lower than that of Portland cement, although its cost differential with respect to limestone was not determined. Bowles pointed out that the slate had a comparable specific gravity with that of limestone but that it had the advantage over Portland cement in providing about 10% more volume for a given weight.

By the early 1930s, slate powder was being used in the United Kingdom as a filler in bituminous road making compositions;[23] slate-filled materials were employed in such applications as sealing and wearing carpets of various kinds and as grouts for setting paving blocks. In a publication in 1962, the United Kingdom Road Research Laboratory provided data relating to slate filler in a discussion of the laboratory design of dense tar surfacing[24] and also pointed out that slate dust had been used satisfactorily as a filler in rolled asphalt.[25]

For many years, bituminous materials have been used as coatings to provide protection for underground pipelines; research in the 1940s resulted in the development of materials that have become accepted and proved in service.[26] Their usefulness in protection against corrosion stems from their chemical inertia, low moisture permeability and water absorption properties; furthermore, they bond well to metals—and their electrical resistance properties are important in connection with the cathodic protection techniques that are often used to reduce corrosion in metals placed underground. Several patents and other publications have referred to the use of slate filler in bituminous pipeline coatings and enamels.[27–32]

There have been a number of examples of uses in other types of coatings. Certain bituminous materials, known as cutback asphalts in American terminology, are used as abrasion-resistant coatings;[33] for these applications, relatively large mineral fibre and filler contents are used to increase the abrasion resistance—and slate powder is one of a number of materials used for this purpose. Bituminous compositions are also used as sound-deadening coatings for metals and provide the additional benefits of increased corrosion resistance, sealing properties, chemical and water resistance and sometimes enable the thickness of the metal to be reduced.[33] The use of these materials began in the motor car industry but has since become much more widespread. Very high filler contents (giving up to 92% solids) are used to provide rigidity and mass and the materials employed include slate powder.

Two patents have described the preparation of bituminous fire-resistant coating materials containing slate powder[34,35] and others have referred to its use in various caulking, jointing, filling, coating and adhesive compositions and putties.[36–39] Apart from aspects of the use of slate as surfacing granules for the manufacture of roofing felts discussed above, there have also been systematic investigations involving its use as an additive and filler in bituminous roofing compositions.[40,41] Finally, an East German patent[42] has described a process for making compression moulded bituminous flooring tiles with slate powder.

2.3.3. Slate powder in plastics

Fillers are used in a large number of plastics; as with bitumens, their addition results in modified properties but, in this case, the cost saving that can be achieved is often particularly significant since the basic polymers are generally relatively expensive. The ideal filler for these materials should be inert and many different minerals, usually in finely ground form, are suitable for this purpose. The colour of the final composition is often important and this can restrict the use of slate powder in these applications. Nevertheless, the dull colouration of slate-filled plastics can, to some extent, be offset with pigments and, in any case, it is not relevant for many technological uses where cost saving and performance characteristics are the predominant factors.

Slate powder has been employed as a filler in plastics for many years. For example, in 1921, Bowles[43] pointed out that it was being used in relatively large quantities for this purpose and one British patent,[44] which dates back over 40 years, referred to the use of resin-bonded slate powder in making roofing slates and similar sheet products.

For some applications, it is possible to use a relatively small proportion of polymeric binder and the quantity of slate used in products of this kind can be quite high. For instance, in an experimental programme aimed at developing synthetic slates and tiles, slate contents of up to 86·5% were achieved.[45] And, in development trials of materials suitable for manufacturing compression moulded synthetic blackboards, polymer contents as low as 6–15% were used.[46] Furthermore, Challis[47] has recently pointed out that polymers with very high loadings of filler can be processed with conventional equipment to give non-brittle products that are both stiff and strong; he predicted that more attention will be paid to such materials although care must be taken in sizing the filler. The need to optimize the relationship between cost and properties of plastics products suggests that applications of this kind, involving relatively high filler contents, could eventually turn out to provide a significant outlet for slate powder—provided that sufficient interest can be generated to promote the necessary development work.

The range of actual applications for which slate filler has already been used or investigated is very wide. The details of many of these remain unpublished, often for commercial reasons, but there are a large number of articles and patents which indicate the variety of possibilities that have been considered. A few of these involve thermoplastics but the majority employ thermosetting resin systems, particularly phenolics, polyesters and epoxies.

Several publications have referred to the use of slate powder as a filler in various thermoplastic hydrocarbon polymers, such as polyethylene and polypropylene,[48-50] and in thermoplastic hot-melt adhesives.[51] Poly(methyl methacrylate) composites containing slate have also been investigated.[52]

Mineral fillers are often used to modify the performance of thermosetting materials[53] and in some applications they can be particularly useful in improving heat resistance and electrical performance. For instance, the use of slate powder as a filler can give resins good antitracking properties;[54] that is to say, it can improve their resistance to the breakdown of electrical insulation performance which results from the surface carbonization of the resin and the consequent formation of a conducting path.

As noted above, phenolic resin systems incorporating slate filler have received some attention in the literature. Wiegand and Müller[55] carried out an investigation into the effect of composition on properties of phenolic moulding materials incorporating a variety of fillers, including slate powder. And Basu[56] investigated the effect of preheating at different temperatures and different periods on the curing time of Novolak resin systems that contained slate and kieselguhr as mineral fillers. Work has also been carried out to develop commercial products based on slate-filled phenolic resins. Two French patents have described such systems for making roofing materials.[57,58] Slate has also been suggested as a filler in the preparation of laminates[59] and it has been used in an investigation into materials for making disc brake pads.[60]

Bossu[61] discussed the formulation of glass-fibre reinforced polyester resin compositions incorporating fillers. He listed a number of different fillers (including slate) and illustrated the variation in cost and various properties with composition using triangular diagrams plotted in terms of resin, fibre and filler content.

In a previously unpublished investigation, Watson, Izod and Riding[62] compared the effect of powdered slate and granite fillers on the flexural properties of a polyester resin. Their test specimens were in the form of beams, supported at each end and subjected to centre-point loading. The particle size distribution of the slate powder was such that 94% by weight passed through a 0·075 mm sieve and 88 % passed through 0·053 mm; and the corresponding quantities for the granite powder were 96 % and 93 % respectively. The filler loadings ranged up to 50 % by weight and Fig. 2.9 shows the effect of filler content on flexural strength, centre-point deflection at failure and flexural modulus. The results are expressed, in relative and non-dimensional terms, as the ratio of values for the filled resin to those of

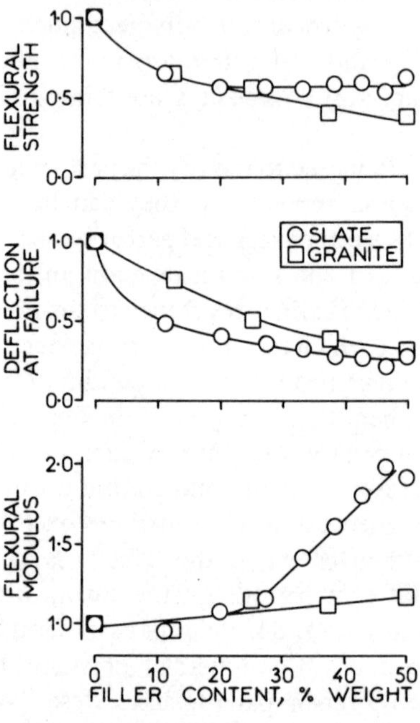

Fig. 2.9. The effect of filler content on the relative values of flexural strength, centre-point deflection at failure and flexural modulus of granite- and slate-filled polyester resin compared with unfilled resin (from ref. 62).

the unfilled resin. The figure shows that the addition of filler leads to reduced flexural strength; increasing filler content, up to about 20 %, leads to a progressively decreasing rate of reduction in flexural strength which, in the case of slate, then levels off up to 50 % filler—on the other hand, the strength of the granite-filled compositions continues to fall at a more or less linear rate with contents above 20 %. The addition of both fillers reduces the centre-point deflection at failure, but the slate powder is the more effective of the two in this respect—it seems to be considerably more effective at lower contents but the deflection values tend to converge as the filler content is increased and there is relatively little difference between the two at 50%. The flexural modulus of the granite-filled material appears to increase more or less linearly over the whole range of filler content; but the slate filler provides a dramatic increase in flexural modulus as its content is raised above about 20–25%. The results indicate that, in the higher range of filler contents, slate powder is the superior of the two; in particular the large increase in modulus with slate content, and the relatively constant strength values in this range, suggest that slate powder may be a useful filler in materials of this kind. There was no systematic attempt to explain the results of this work but it is suspected that the shape of the filler particles may be at least partly responsible for the differences that were observed.

As with phenolic resins, there has been some commercial development of slate-filled polyester systems. A variety of decorative effects can be achieved by means of powdered fillers and granules dispersed in cast resins and also by embedding stone chips or slices in the surfaces of such materials. The techniques employed have been discussed by Winfield[63] and decorative materials of this kind have potential in architectural applications as well as in the manufacture of products for interior decoration and furnishing. As another example, chess pieces have been produced from a 50/50 mixture of polyester resin and slate powder with the addition of a small quantity of pigment to help offset the dull colour of the slate.

The range of such products that can be manufactured is obviously enormous although, as Winfield pointed out, little detailed information of individual production methods is generally available. The scale of manufacture can vary greatly; depending on the nature of the product, the cost of setting up modest production facilities (for producing small items such as chess sets, for example) can be low and manufacture is sometimes carried out on a very small scale. On the other hand, the mass production of architectural components can involve considerable capital investment.

There have been several references to building and constructional applications for slate-filled polyester resins, for instance in making floor tiles,[64] resin-bonded 'concrete',[65] and for fixing anchor bars in drilled holes[66] as well as for other adhesives.[67]

Fillers are mixed with epoxy resins to reduce the cost of the material and to modify properties—particularly curing shrinkage, compressive strength and thermal characteristics.[53] Slate powder is a commonly used filler for this purpose and Table 2.2, derived from data published by Lannon and Hoskins,[53] shows the relative effect of the addition of approximately 20% slate filler on a number of the properties of an epoxy resin; the figures given are the ratios obtained by dividing the values for the filled material by those for the unfilled resin. The improved compressive strength can be seen to be accompanied by a loss in both flexural and impact strength. Furthermore, the thermal conductivity is increased and both the thermal expansion and water absorption are reduced by the presence of the slate filler. (Lannon and Hoskins noted that increased thermal conductivity, which facilitates the dissipation of heat, reduces the harmful effect of the internal temperature

Table 2.2. The relative effect of the addition of approximately 20% slate filler on the properties of an epoxy resin (after Lannon and Hoskins[53]).

Property	*Relative values for the filled material compared with those for the unfilled resin*
Density	1·36
Compressive strength	1·22
Flexural strength	0·63
Impact strength	0·72
Thermal conductivity	1·78
Thermal expansion	0·82
Water absorption	0·67

rise of the material due to the exotherm which accompanies the curing of the resin, particularly when thick sections are used.)

A number of commercial applications for slate-filled epoxy resins can be found in the literature, including abrasion-resistant coatings[68] and compositions for repairing bridges.[69]

There are also a number of miscellaneous applications, that are based on other polymer systems, which include vibration- and sound-absorbing materials[70,71] and chemically resistant compositions.[72]

The range of publications suggests that there has been a considerable amount of *ad hoc* development work in connection with specific applications, but it seems that there has probably been less attention paid to broader investigations than in the case of bituminous compositions. In view of the enormous variety of plastics available, and their wide range of properties and applications, this is perhaps not surprising. Nevertheless, there are gaps which need to be filled. For instance, it has been noted that there can be problems in 'wetting' slate filler with polymers, particularly at low binder contents;[3] and a further investigation of this aspect of its behaviour might lead to improved properties. There also appears to be a need for more comparative data to assess the performance of slate powder in relation to other more widely used mineral fillers.

2.3.4. Slate powder in rubber

In 1921, Bowles[43] described one of the earliest systematic investigations into the use of slate powder as a filler in rubber; this was the result of the co-operation of a number of rubber product manufacturers with the United States Bureau of Mines and involved the evaluation of slate powder samples which mostly contained 80% material passing through a 200 mesh screen. Of the 27 companies supplied with samples, seven replied with adverse reports claiming that the slate-filled rubbers had inferior properties which included tensile strength and hardness as well as stretch, wearing and cracking characteristics—and several reports suggested that the results would have been better if the material had been more finely ground. One set of tests compared the slate filler with soapstone and showed that the slate gave considerably lower tensile strength and elongation values, but the slate grains were coarser and less uniform than those of the soapstone. Eleven of the manufacturers submitted favourable reports and found that the slate powder was a satisfactory filler for a variety of products including moulded shoe soles and heels, battery jars and garden hose—and it was successfully substituted for whiting, clay, ground barytes and aluminium flake. A footwear manufacturer found that slate-filled rubber compared favourably with conventional materials in its tensile strength, elongation and aging properties but, on repeated stretching, it failed more readily—and it was suggested that this difficulty would have been overcome by finer grinding of the powder. Another set of tests gave good results in comparing slate powder with whiting and clay fillers on the basis of tensile strength and elongation properties; and it was concluded that slate could be used advantageously if the price was low enough. Furthermore, a manufacturer of mechanical rubber goods, tubing and tape found that slate powder could be used instead of cheap whiting in grey and black rubber products.

One slate sample was pulverized so that 95% passed a 200 mesh screen and 90% passed 300 mesh—and tests showed that this material gave results that were superior to those obtained with whiting or ground barytes but slightly inferior to those with high grade clay.

Bowles suggested that the conflicting opinions may have stemmed largely from the inadequate grinding of the material and that the evidence pointed to the probability that fine slate powder is a satisfactory filler for mechanical rubber goods—and he noted the price advantage of slate powder compared with conventional fillers.

As part of the same investigation, some tests were made with slate powder in rubber compounds of the kind that were used for motor car tyres at that time but, in this application, its performance was found to be inferior to that of the fillers normally used for this purpose.

There have been a few other relevant publications[73-75] since Bowles' report but it appears that there is little if any significant use of slate powder by the rubber industry at present.

It has been said that certain slates contain constituents which can lead to the deterioration of rubber into which they are incorporated;[45] this possibility should, of course, be considered if the use of slate filler is being contemplated in rubber compounds.

2.3.5. Slate powder in paints and similar surface coatings

It has been pointed out that both quartz and mica, which are major constituents of slate, have been used individually as fillers or extenders in paints;[3] quartz has been added to improve sanding properties and to enhance the abrasion resistance of the paint film—and mica has been used as a suspension aid, and also to strengthen the paint film and increase its resistance to moisture, as well as to dilute primary pigments in the manufacture of inexpensive paints. This suggests that slate powder may be a useful ingredient in formulating paints.

In fact, slate powder has been used by the paint and allied industries for at least 50 years; a number of publications in the 1920s and 1930s,[76-78] and both during and shortly after the Second World War,[3,79-82] referred to its application in paints and similar protective coatings. Since that time, references in the literature have tended to be less specific, simply referring to the general use of slate powder in paints as one of a number of applications.

Although the powder is fairly cheap, its current consumption by the paint industry is probably relatively small in comparison with other materials used for similar purposes; the total annual consumption of slate powder for paint manufacture in the United Kingdom is probably less than 100 tonnes.[83] It is used mainly as an extender pigment and filler rather than as a primary pigment for colour or anticorrosive properties. It is particularly useful in refinishing paints for the motor vehicle repair trade where the film levelling properties and the rubbing down and flatting characteristics, which it imparts to the film, are important. The extent to which it is used in the finishes is limited mainly by colour, which is that of the slate when the powder is dispersed in water, and can be a dark shade of chocolate when the powder is dispersed in oil; it is therefore more commonly used in primers and surfacers where colour is of less importance.

It has been suggested that the use of slate powder as a pigment is a solution to the problem of 'marine chalking'.[84] In this context, the term 'chalking' refers to the development of a loose powder, on a painted surface, that results from the deterioration of the paint film because of weathering; the problem can be particularly serious in a marine environment.

2.3.6. Slate powder in fertilizers, insecticides and pesticides

Apart from the use of slate granules mentioned earlier, a number of other applications have been discussed in connection with the agricultural and related industries. These have ranged from the addition of slate to improve soil[85,86] to its use as a component in fertilizer compositions.[87,88] There have also been a number of references to the use of slate powder in insecticide and pesticide compositions.[89–91]

2.3.7. Slate powder in cosmetic preparations

A number of publications have referred to the use of slate powder in formulating cosmetic preparations such as powders and dentifrices. In view of the health hazard known to be associated with slate powder, and the availability of more suitable materials for such applications, it seems most unlikely that it is now used by any manufacturers of products of this kind. (There has been at least one instance in the past in which the incorporation of slate powder into a cosmetic preparation has led to harmful effects in use.)

2.3.8. Miscellaneous uses

Apart from the actual and potential uses already discussed, there have been a number of references describing miscellaneous applications which do not readily fall into the categories above.

Earlier sections have referred to a range of materials made from slate powder using organic binders of various kinds. From time to time, patents have also been published which describe materials that are essentially inorganic in character[92–99] using a variety of binders including hydraulic cements. There have also been references to using slate in making cardboard[100] and paper[101] and in formulating dry fire-extinguishing powders.[102] It has also been used for glazing fired clay products;[3] a suspension of slate powder in water, applied to clay, was found to give a brilliant brown glaze on firing.

REFERENCES

1. R. B. Lewis, 'Commercial aspects of asphalts', *Bituminous Materials: Asphalts, Tars, and Pitches* (Editor: A. J. Hoiberg), Volume II: Asphalts, Part I, Interscience Publishers, New York (1965), 123–49.
2. G. W. Berry, 'Roofing materials', *Kirk–Othmer Encyclopedia of Chemical Technology*, 2nd Edition, Volume 17, Interscience Publishers, New York (1968), 459–75.
3. *Properties and New Uses of Pennsylvania Slate*, Mineral Industries Experiment Station Bulletin No. 47, Pennsylvania State College (1947).
4. C. L. Jewett and R. C. Collins, 'Construction materials—granules', *Industrial Minerals and Rocks (Nonmetallics other than Fuels)*, 4th Edition (Editor-in-Chief: S. J. Lefond), American Institute of Mining, Metallurgical, and Petroleum Engineers, Inc., New York (1975), 175–84.
5. 'Penrhyn expands production of slate powder and granules', *Industrial Minerals*, No. 40 (Jan. 1971), 37–41.
6. Czech. 146,793 (1972). (*Chem. Abstr.*, **78** (1973), 139724t.)
7. Czech. 160,195 (1975). (*Chem. Abstr.*, **84** (1976), 168809d.)
8. B. Sumner, 'Minerals as granular carriers for pesticides', *Proceedings of the First 'Industrial Minerals' International Congress, London*, 8–9 July 1974, Metal Bulletin Ltd, London (1975), 100–3.
9. US 3,137,618 (1964). (*Chem. Abstr.*, **62** (1965), 2203f.)
10. R. N. Crockett, *Slate*, Mineral Dossier No. 12, Mineral Resources Consultative Committee, HMSO, London (1975).
11. E. J. Humby (Department of Civil Engineering, Portsmouth Polytechnic), private communication.
12. '*Delafila*' *Slate Powder*, Information Sheet No. 4A-1, Delabole Slate (1977) Ltd, Cornwall.
13. R. N. J. Saal, 'Asphalt systems', *Composite Materials* (Editor: L. Holliday), Elsevier Publishing Company, Amsterdam (1966), 453–74.
14. P. B. Shepherd, 'Mineral fillers', *Bituminous Materials: Asphalts, Tars, and Pitches* (Editor: A. J. Hoiberg), Volume I, Interscience Publishers, New York (1964), 347–73.
15. *Bituminous Materials in Road Construction*, Road Research Laboratory, Department of Scientific and Industrial Research, HMSO, London (1962), 146–7.
16. N. Severinghaus, Jr, 'Fillers, filters, and absorbents', *Industrial Minerals and Rocks (Nonmetallics other than Fuels)*, 4th Edition (Editor-in-Chief: S. J. Lefond), American Institute of Mining, Metallurgical, and Petroleum Engineers, Inc., New York (1975), 235–49.
17. H. Kunath, 'Properties of bituminous mortars with activated filler', *Proceedings of the Second International Symposium on Tests on Bitumens and Bituminous Materials, Budapest*, September 1975, 449–62.
18. H. Walther, *Kolloid Zeitschrift*, **103** (1943), 57.
19. S. H. Alexander and G. W. Tarver, 'Permeability and water absorption', *Bituminous Materials: Asphalts, Tars, and Pitches* (Editor: A. J. Hoiberg), Volume II: Asphalts, Part I, Interscience Publishers, New York (1965), 223–47.

20. H. W. Slotboom, in *The Properties of Asphaltic Bitumen* (Editor: J. Ph. Pfeiffer), Elsevier, New York (1950), 270, 272.
21. Ref. 18, 59.
22. O. Bowles, *Slate Dust in Asphalt Road Surface Mixtures*, Bureau of Mines Report of Investigations No. 2230, United States Department of the Interior (Mar. 1921).
23. 'The utilization of slate waste—how waste quarried slate now plays an important part in modern road construction', *The Quarry and Roadmaking*, **36** (407) (Jan. 1931), 12–13.
24. Ref. 15, 275–81.
25. ibid., 266.
26. G. W. Tarver and S. H. Alexander, 'Asphalt pipe coatings', *Bituminous Materials: Asphalts, Tars, and Pitches* (Editor: A. J. Hoiberg), Volume II: Asphalts, Part I, Interscience Publishers, New York (1965), 479–510.
27. A. Payer, *Hutn. Listy*, **28**(2) (1973), 273–5. (*Chem. Abstr.*, **79** (1973), 106978w.)
28. US 2,767,102 (1956). (*Chem. Abstr.*, **51** (1957), 1625i.)
29. Fr. 1,154,869 (1958). (*Chem. Abstr.*, **55** (1961), 7867f.)
30. US 2,973,280 (1961). (*Chem. Abstr.*, **55** (1961), 14941c.)
31. Ger. 1,167,727 (1964). (*Chem. Abstr.*, **68** (1968), 4685d.)
32. Fr. 1,336,685 (1963). (*Chem. Abstr.*, **60** (1964), 765a.)
33. C. C. Weeks, 'Industrial protective coatings', *Bituminous Materials: Asphalts, Tars, and Pitches* (Editor: A. J. Hoiberg), Volume II: Asphalts, Part I, Interscience Publishers, New York (1965), 511–57.
34. Ger. 1,279,262 (1968). (*Chem. Abstr.*, **70** (1969), 30179c.)
35. Ger. Offen. 2,400,888 (1975). (*Chem. Abstr.*, **84** (1976), 34697g.)
36. US 2,675,365 (1954). (*Chem. Abstr.*, **48** (1954), 10261d.)
37. Ger. 1,242,138 (1967). (*Chem. Abstr.*, **67** (1967), 66359p.)
38. East Ger. 38,056 (1965). (*Chem. Abstr.*, **63** (1965), 10178b.)
39. Brit. 1,125,518 (1968). (*Chem. Abstr.*, **69** (1968), 79073g.)
40. S. H. Greenfeld, *Natl Bur. Standards, Bldg Materials Structures Rept*, No. 147 (1956). (*Chem. Abstr.*, **51** (1957), 3126a.)
41. K. G. Martin, *CSIRO, Australia, Div. Building Res. Tech. Paper*, No. 12 (1962). (*Chem. Abstr.*, **56** (1962), 14530a.)
42. East Ger. 74,468 (1970). (*Chem. Abstr.*, **74** (1971), 114081k.)
43. O. Bowles, *The Utilization of Waste Slate as a Filler*, Bureau of Mines Report of Investigations No. 2283, United States Department of the Interior (Sep. 1921).
44. Brit. 497,062 (1938). (*Chem. Abstr.*, **33** (1939), 3559[1].)
45. R. H. Boyle (Director, Penrhyn Quarries Ltd), private communication.
46. E. C. Henry, 'Utilization of waste slate', *J. Can. Ceram. Soc.*, **17** (1948), 27–31.
47. A. A. L. Challis, 'Polymer engineering', *Chemistry in Britain*, **14**(9) (Sep. 1978), 446, 449, 450, 452, 454.
48. J. Hladik, *Plasty Kauc.*, **11**(8) (1974), 239–41. (*Chem. Abstr.*, **83** (1975), 115480n.)
49. Ger. Offen. 2,423,128 (1974). (*Chem. Abstr.*, **82** (1975), 99589h.)
50. Brit. 1,134,422 (1968). (*RAPRA Abstr.*, **3** (Apr. 1969), 3102P.)
51. East Ger. 60,607 (1968). (*Chem. Abstr.*, **70** (1969), 38459q.)

52. H. C. Pyun, B. R. Cho and S. K. Kwon, *J. Korean Nucl. Soc.*, **7**(1) (1975), 9–14. (*Chem. Abstr.*, **84** (1976), 45467h.)
53. D. A. Lannon and E. J. Hoskins, 'Effect of plasticisers, fillers, and other additives on physical properties', *Physics of Plastics* (Editor: P. D. Ritchie), The Plastics Institute, London, Illiffe Books Ltd (1965), 323–86.
54. K. L. Loewenstein, 'Glass systems', *Composite Materials* (Editor: L. Holliday), Elsevier Publishing Company, Amsterdam (1966), 129–220.
55. H. Wiegand and K. Müller, *Kunstst. Rundsch.*, **16** (1969): (10) (Oct.), 595–603; (11) (Nov.), 662–5; (12) (Dec.), 716–20. (*Chem. Abstr.*, **72** (1970), 101321x.)
56. D. P. Basu, *J. Inst. Chemists (India)*, **39**(5), (Sep. 1967), 195–8. (*Chem. Abstr.*, **68** (1968), 78992y.)
57. Fr. 1,464,845 (1967). (*Chem. Abstr.*, **67** (1967), 46912k.)
58. Fr. Demande 2,288,610 (1976). (*Chem. Abstr.*, **86** (1977), 17973t.)
59. Belg. 641,077 (1964). (*Chem. Abstr.*, **63** (1965), 4508d.)
60. T. Jakubiuk, J. Janecki and S. Zawalski, *Polimery (Warsaw)*, **20**(3) (1975), 145–6. (*Chem. Abstr.*, **84** (1976), 5801f.)
61. B. Bossu, 'Expériences sur les compounds fibre de verre-polyester', *Industrie des Plastiques Modernes*, **10**(7) (Jul.–Aug. 1958), 27–30. (*Chem. Abstr.*, **53** (1959), 1810g.)
62. K. L. Watson, D. A. Izod and C. P. Riding (Department of Civil Engineering, Portsmouth Polytechnic), unpublished work.
63. A. G. Winfield, 'Cultured stones—decorative embedments in plastics', *Soc. Plastics Engrs J.*, **22**(8) (Aug. 1966), 36–9.
64. Fr. 1,561,500 (1969). (*Chem. Abstr.*, **71** (1969), 82319p.)
65. S. African 6,805,597 (1970). (*Chem. Abstr.*, **73** (1970), 69510r.)
66. Fr. 1,531,955 (1968). (*Chem. Abstr.*, **71** (1969), 31147y.)
67. K. H. Kuessner, *Plaste u. Kautschuk*, **7** (1960), 431–2. (*Chem. Abstr.*, **55** (1961), 14974h.)
68. Czech. 128,324 (1968). (*Chem. Abstr.*, **71** (1969), 61977m.)
69. W. Franke, 'The employment of epoxy resin for the restoration of bridge members', *RILEM Symposium on Experimental Research on New Developments brought by Synthetic Resins to Building Techniques, Paris*, 4–6 September 1967. (Also *Chem. Abstr.*, **76** (1972), 86733v.)
70. S. V. Martirosyan and I. A. Saakyan, *Prom-st. Arm.* (4), (1975), 62–6. (*Chem. Abstr.*, **83** (1975), 61856u.)
71. Fr. 1,359,183 (1964). (*Chem. Abstr.*, **62** (1965), 2893d.)
72. US 2,653,920 (1953). (*Chem. Abstr.*, **48** (1954), 975c.)
73. J. R. Scott, *J. Research Assoc. Brit. Rubber Mfrs*, **1** (1932), 115–26. (*Chem. Abstr.*, **27** (1933), 4127.)
74. J. F. Sacher, *Gummi-Ztg.*, **47** (1933), 439. (*Chem. Abstr.*, **27** (1933), 1783.)
75. T. R. Dawson, *J. Rubber Research*, **14** (1945), 8–10. (*Chem. Abstr.*, **39** (1945), 2005^4.)
76. *Peintures, Pigments, Vernis*, **3** (1926), 4–9. (*Chem. Abstr.*, **20** (1926), 2254.)
77. H. Rabaté, *Peintures, Pigments, Vernis*, **9** (1932), 1–9. (*Chem. Abstr.*, **26** (1932), 2606.)
78. H. Rabaté, *Industrie Chimique*, **23** (1936), 486–90, 663–7. (*Chem. Abstr.*, **32** (1938), 3510^5.)
79. C. L. Harness, '*Natural Mineral Paint Extenders*', Bureau of Mines

Information Circular 7264, United States Department of the Interior (1943). (*Chem. Abstr.*, **38** (1944), 1382[6].)
80. *Paint and Varnish Production Manager*, **22** (1942), 7–8, 10. (*Chem. Abstr.*, **36** (1942), 1504[9].)
81. F. J. North, *The Slates of Wales*, 3rd Edition, The National Museum of Wales (1946), 83.
82. U. Cuppini, *Pitture e Vernici*, **4** (1948), 475–7. (*Chem. Abstr.*, **44** (1950), 1370f.)
83. D. G. Dawson (Paints Division, Imperial Chemical Industries Ltd), private communication.
84. *Paint Film Defects, Their Causes and Cure* (Editor: M. Hess), 2nd Edition, Chapman and Hall Ltd, London (1965), 427.
85. J. Kobus, *Pamiet. Pulawski*, (63) (1975), 143–62. (*Chem. Abstr.*, **85** (1976), 31954y.)
86. N. G. Zhezel, *Zapiski Leningrad. Sel'skokhoz. Inst.*, No. 11 (1956), 192–202. (*Chem. Abstr.*, **53** (1959), 1615g.)
87. W. Trzesniowski, E. Klinger and B. Rychlik, *Przem. Chem.*, **54**(11) (1975), 643–5. (*Chem. Abstr.*, **84** (1976), 16157b.)
88. Ger. 973,050 (1959). (*Chem. Abstr.*, **55** (1961), 2984c.)
89. W. Maldwyn Davies, *Welsh J. Agr.*, **13** (1937), 317–20. (*Chem. Abstr.*, **31** (1937), 8800[6].)
90. W. A. L. David and B. O. C. Gardiner, *Bull. Entomol. Research*, **41** (1950), 1–61. (*Chem. Abstr.*, **44** (1950), 10998b.)
91. M. L. Turich, *Gigiena i Sanit.* (2) (1955), 50–1. (*Chem. Abstr.*, **49** (1955), 14266d.)
92. Ger. 524,774 (1927). (*Chem. Abstr.*, **25** (1931), 4380.)
93. Brit. 355,056 (1930). (*Chem. Abstr.*, **27** (1933), 396.)
94. US 1,362,563 (1920). (*Chem. Abstr.*, **15** (1921), 582.)
95. Russ. 36,261 (1934). (*Chem. Abstr.*, **32** (1938), 5376[1].)
96. US 2,785,987 (1957). (*Chem. Abstr.*, **52** (1958), 1584b.)
97. Brit. 958,174 (1964). (*Chem. Abstr.*, **61** (1964), 1596e.)
98. Brit. 257,115 (1925). (*Chem. Abstr.*, **21** (1927), 3109.)
99. Brit. 949,401 (1964). (*Chem. Abstr.*, **66** (1967), 5500a.)
100. Russ. 8,630 (1926). (*Chem. Abstr.*, **25** (1931), 1048.)
101. A. Esme, *Papeterie*, **60** (1938), 350, 353. (*Chem. Abstr.*, **32** (1938), 5626[3].)
102. Ger. 1,131,999 (1962). (*Chem. Abstr.*, **58** (1963), 12208d.)

3

The Use of Slate Waste in Concrete

3.1. INTRODUCTION

The literature contains references to various possibilities for using slate in concrete and there has been significant utilization in this application, particularly in lightweight concrete.

As noted in Chapter 1, slate, because of its geological history, has a chemical composition which is similar to that of many clays. This fact has suggested its use for the production of Portland cement and, for instance, Chapter 1 refers to a scheme for cement manufacture that was discussed as far back as 1895. It has been pointed out that, as far as England and Wales are concerned, slate waste is not located conveniently close to limestone sources;[1] however, it was considered, some years ago, for cement manufacture in Scotland but the site for a cement works would depend on the location of the limestone and the relative position of potential market areas rather than the source of argillaceous material.[2] Furthermore, the slate would require grinding.[1] Nevertheless, this application has been considered elsewhere.[3]

The great majority of publications referring to the use of slate in concrete are concerned with expanded slate as a lightweight aggregate, and this will be the major topic of discussion in the present chapter. But slate powder has also been investigated as a concrete 'admixture', and some use has been made of crushed, unexpanded slate as a dense aggregate; these applications will be discussed first.

3.2. SLATE POWDER AS A CONCRETE 'ADMIXTURE'

Slate powder has been considered as a partial replacement for cement in conventional concrete. Watson, Grant and Jones[4] determined the compressive strength of water-saturated paste specimens, containing no aggregate, in which fine slate powder (94% passing through a 0·075 mm sieve and 88% passing 0·053 mm) had been blended with ordinary Portland cement using 0, 10 and 20% slate in the dry mix; water/total solids ratios of 0·35 and 0·45 were used and the specimens were cured under water for varying periods up to 28 days. Specimens were also prepared from corresponding blends using fine granite powder (96% passing 0·075 mm

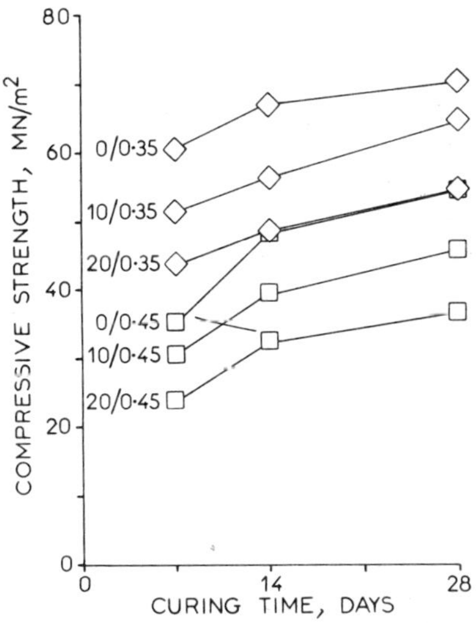

Fig. 3.1. The development of compressive strength with curing time for water-cured cement/slate powder mixes with replacement levels of 0, 10 and 20% slate and water/solids ratios of 0·35 and 0·45 (from ref 4)

and 93% passing 0·053 mm) then cured and tested in the same way. The results using slate powder, which are plotted in Fig. 3.1, show that the 0% slate/0·45 water/solids mix (with, of course, a water/cement ratio of 0·45) has virtually the same strength as the 20% slate/0·35 water/solids mix at 14 and 28 days; the water/cement ratio of the latter mix, assuming the slate powder to be inert and non-absorbent, is 0·44 which is almost the same as that of the former. Thus the equal strengths suggest that the slate powder may act as inert filler that merely serves to raise the water/cement ratio above the water/total solids ratio to an extent that depends directly on its content. The specimens prepared from the granite powder gave similar results for the corresponding mixes.

The effect of partial replacement was established by plotting the compressive strength of all the specimens against their respective cement/water ratios at the different curing times employed. These results, and others, that were obtained after 28 days' curing are shown in Fig. 3.2. After both 14 and 28 days' curing, the plots showed that there is a linear increase in compressive strength with cement/water ratio (and the linear correlation coefficients are very high); furthermore, the established fact that the compressive strength of normal concrete generally varies in this way[5]

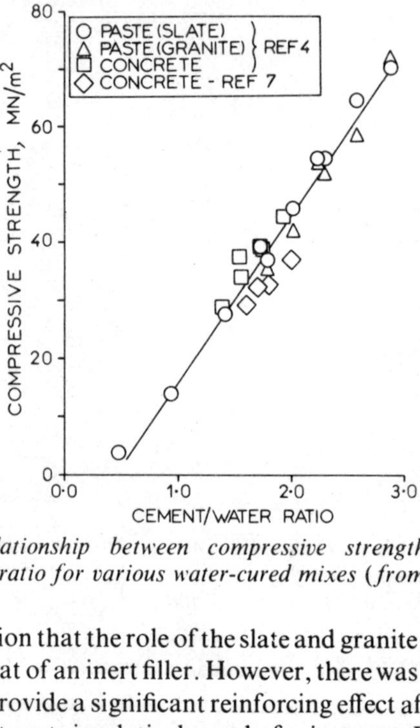

Fig. 3.2. The relationship between compressive strength at 28 days and cement/water ratio for various water-cured mixes (from refs. 4 and 7).

led to the conclusion that the role of the slate and granite powders does seem to be primarily that of an inert filler. However, there was some evidence that both fillers may provide a significant reinforcing effect after 7 days when the hardening cement paste is relatively weak; for instance, Fig. 3.1 shows that, at this age, the 20 % slate/0·35 water/total solids mix is considerably stronger than the 0 %/0·45 mix.

Unpublished work[6] showed that slate powder/Portland cement blends which had been steam cured for 6 hours (at 96–98 °C at atmospheric pressure after precuring for 1, 7 and 14 days) also gave linear plots of strength against cement/water ratio that suggested the inert nature of the slate powder under these curing conditions; some of these plots are shown in Fig. 3.3. (Results obtained under autoclaving, i.e. high-pressure steam curing, conditions are quite different and will be discussed in Chapter 5.)

Watson, Grant and Jones[4] extended their work with normally cured materials to include data from concrete specimens; again they used cement replacement levels of 0, 10 and 20% slate but with an aggregate/(cement + slate) ratio of 5·04 and water/(cement + slate) ratios of 0·52 and 0·58. These materials were cured under water for 28 days before compression testing. And, as Fig. 3.2 shows, the results fell very close to the plot obtained with the 28 day old paste specimens prepared with no aggregate;

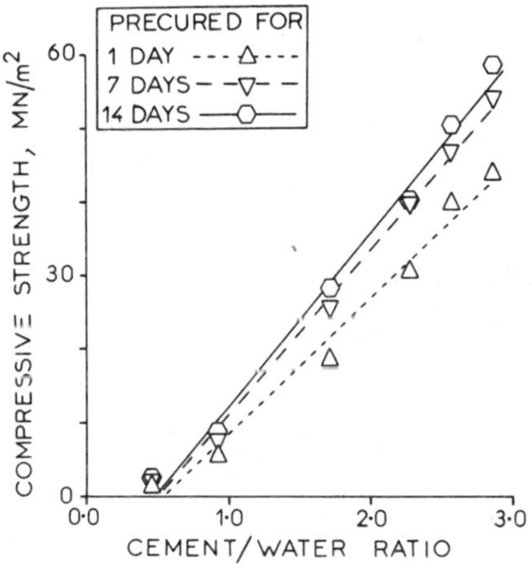

Fig. 3.3. The relationship between compressive strength and cement/water ratio for cement/slate powder mixes that were steam cured, after precuring for various periods, and tested on the following day (from ref. 6).

furthermore, statistical analysis showed that the combined results gave a regression line that was virtually the same as that obtained with the paste results alone.

The workability of these concrete mixes, when fresh, was assessed by using slump, compacting factor and Vebe tests, and the results are shown in Table 3.1. The data, although limited, suggest that the replacement of cement by slate powder, for the mixes used, has little effect although there is some indication that the 10% replacement level may improve the workability, particularly for the drier mix.

An independent and unpublished investigation[7] was carried out to assess the pozzolanic activity of slate powder using a control mix having a water/cement ratio of 0·50 and an aggregate/cement ratio of 5·49; three further mixes were prepared in which 10, 15 and 20% of the cement was replaced by slate powder. Compressive strength was determined after 3, 7

Table 3.1. The workability of some concrete mixes with partial replacement of the cement by slate powder (from ref. 4).

Water/solids ratio	*Slate replacement* (%)	*Slump* (*mm*)	*Compacting factor*	*Vebe* (*s*)
0·52	0	13	0·92	7
	10	32	0·92	6
	20	13	0·88	8
0·58	0	64	0·94	3
	10	70	0·96	3
	20	57	0·95	3

and 28 days' water curing. Figure 3.4 shows that the results of this work also demonstrated a more or less linear increase in compressive strength with cement/water ratio which suggested that the slate powder acts as an inert filler; and Fig. 3.2 shows that the 28 day values fall very close to the plot obtained by Watson, Grant and Jones. Furthermore, in this case, the addition of 3% slate powder (by weight of cement) to the control mix was found to increase the slump value from 13 to 29 mm and gave the concrete a more 'binding' consistency; but it should be remembered that the aggregate/cement ratio was higher and the water/cement ratio lower than those used in the other investigation.[4]

Watson, Grant and Jones concluded that the partial replacement of Portland cement by slate powder in ordinary concrete does not seem to present exciting prospects;[4] admittedly there is the possibility of cost reduction, and perhaps increased workability, provided that the lower compressive strengths can be tolerated—but data is very limited and much

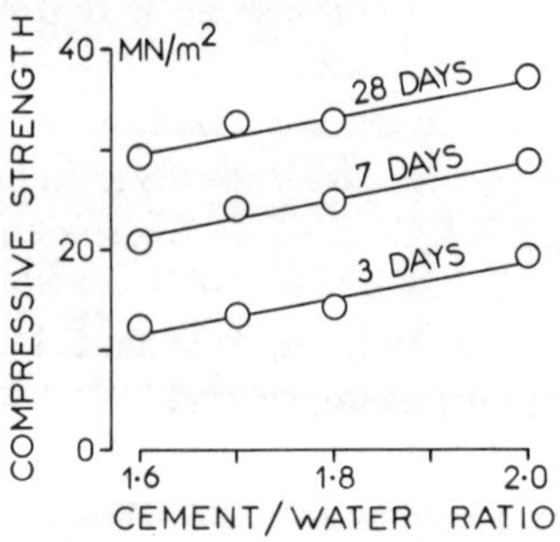

Fig. 3.4. The relationship between compressive strength and cement/water ratio for concrete mixes, with partial replacement by slate powder, at various ages (from ref. 7).

more work would be needed to ensure that other properties are not degraded to an unacceptable level.

(It is perhaps relevant, in this context, to note that slate clay has been considered as an additive in refractory concrete.[8])

3.3. CRUSHED SLATE AS DENSE AGGREGATE IN CONCRETE

A number of publications[1,9,10] have referred to the possibility of using slate as a dense aggregate in concrete for relatively undemanding applications in the United Kingdom; but, of course, the economic distance that the material can be transported is small and such uses are generally likely to be of interest only in areas that are located relatively close to sources of slate waste.

The slate industry in the United Kingdom has long been aware of the possibility of manufacturing precast concrete products from its waste. In 1933, crushed and screened material, from 13 mm size down to dust, was being used for the manufacture of blocks at a quarry in Cornwall in England;[11] the blocks measured 0·46 m by 0·23 m in various thicknesses up to 0·15 m and the plant's capacity was about 1000 blocks per day. A concrete brick making plant, using slate waste, was installed in Scotland in 1954;[1] the bricks satisfied the appropriate British Standard requirements (Class B, BS 1180:1944) but were commercially unsuccessful due to the competition offered by granite-based blocks that were more attractive. (The same reference[1] also mentions the development of an experimental slate-based refractory brick made from high alumina cement.)

In 1956, Hoyt described investigations, carried out in the United States, which included the preparation of bricks from crushed, screened slate;[12] the cement/sand/slate aggregate ratio (using air-entraining cement) was maintained at 1:1:5, although one specimen was prepared without sand using a cement/aggregate ratio of 1:8. The bricks were hand moulded and had an average density of about 1695 kg/m^3; flexural strengths ranged between 4·2 and 8·0 MN/m^2, and compressive strengths between 10·3 and 19·4 MN/m^2 with an average value of 15·3 MN/m^2. The most important single factor was thought to be the particle size distribution of the aggregate and the sieve analysis for the material discussed by Hoyt is shown in Table 3.2. He suggested that, by controlling this factor, the strength could be rigidly controlled and the use of air-entraining agents could often be avoided. He also discussed the equipment required for slate brick manufacture, and its cost, as well as alternative processes that could be used.

By 1959, concrete bricks and tiles were being produced at the Dinorwic Quarry in North Wales.[13] The tile plant had been newly installed and, at that time, was believed to be unique in using slate aggregate. Brick making had begun in 1954 and the plant was producing 24 000–27 000 bricks per shift and was operated for two shifts per day. The primary jaw crusher in this plant was run with the jaws fully loaded since trickle feeding tended to produce elongated slivers of material. The bricks were steam cured for 8 hours.

Table 3.2. Sieve analysis of slate used as aggregate in concrete bricks (from ref. 12).

Sieve size (mm)	*Weight (%)*
Above 18·86	7·88
18·86–4·70	39·64
4·70–3·33	9·81
3·33–1·67	15·92
1·67–0·83	13·54
0·83–0·59	4·36
0·59–0·42	2·34
0·42–0·29	1·04
0·29–0·21	1·53
0·21–0·15	1·40
0·15–0·07	1·30
Below 0·07	1·24

Concrete facing bricks, with ordinary Portland cement as the binder, are currently produced in North Wales using crushed and graded slate (conforming to BS 882, 1201) as the aggregate.[14] These bricks are made on fully automatic presses ensuring precision with clean, sharp arrises and a deep frog. They have a distinctive texture and are supplied in either 'natural' or coloured varieties, the pigment being used to subdue the tendency for the cement to detract from the slate's natural colour. The bricks are in accordance with BS 1180:1872 and fall into the 15 MN/m^2 strength category; their drying shrinkage of 0·026 % (from saturated to oven dry under specified conditions) is well below the maximum of 0·04 % allowed by the standard. They may be used in most types of domestic and industrial building in a variety of applications for wall construction.

Apart from the development of processes for the manufacture of precast concrete products under factory conditions, there has been some work of more general interest—and slate has been used in concrete for more substantial applications. There are very few publications concerned with this topic and it is worth considering the available sources of information in some detail.

In 1966, McGhee and Ozol reported the results of an evaluation of one specific slate as an aggregate in concrete;[15] this work, initiated by the Virginia Highway Research Council, had been carried out as a feasibility study into extending the use of slate in Portland cement concrete in an area without recognized sources of good quality aggregates.

Table 3.3. Selected concrete compositions using slate as dense aggregate (from ref. 15).

Water/cement ratio	0·49	0·44	0·44[a]	0·41[a]
Cement content (kg/m³)	362	376	376	376
Cement/aggregate ratios	1:1·83:3·07	1:1·72:2·93	1:1·72:2·93	1:1·72:2·93
Air content (%)	5·6	5·0	6·4	4·5
Slump (mm)	48	33	71	33

[a] Water-reducing admixture added.

McGhee and Ozol pointed out that the platey and often elongated shape associated with slate particles is considered to be unconducive to good workability and may have an adverse effect upon strength and durability; but their review of the subject included references indicating that the relevant literature, at that time, had apparently revealed no instances in which the use of flakey, elongated materials had led to actual failures and, furthermore, that such materials could be used to prepare concrete of adequate workability. With this in mind, they prepared various concretes using a coarse aggregate made from a crushed slate (graded to conform to an appropriate Virginia Department of Highways specification for concrete aggregate) with a fineness modulus of 7·90; silica sand, with a fineness modulus of 2·83, was used as the fine aggregate. An air content of about 6 % was achieved using a quantity of vinsol resin air-entraining agent that was kept constant in all the mixes. A Type II Portland cement was used which had an alkali content considerably higher than that recommended for use with aggregates subject to alkali silica reactions. Trial mixes were used to achieve adequate workability and the compositions selected for testing are shown in Table 3.3. The addition of a water-reducing admixture was found to increase the slump and the ease of finishing the concrete; it also enabled a reduction in the water content to be achieved, for an equivalent slump to that of concrete with no admixture, and laboratory observations indicated that workability was still markedly improved.

The results of 28 day strength tests and the dynamic elastic properties are shown in Table 3.4; two slate-based concretes are compared with mixes based on a foliated micaceous marble, which had sometimes shown poor performance in concrete, and a granite gneiss which had performed well. The second slate-based concrete listed and the marble- and granite-based concretes were all mixed using similar proportions (see the first column of Table 3.3); the flexural and indirect tensile strengths and the dynamic elastic properties of these three materials are very similar but the compressive strengths differ considerably. McGhee and Ozol pointed out that axial compressive failure can be interpreted in terms of combined shearing and normal stresses determined by the internal friction and cohesive characteristics of the material; but in axial tension, only tensile stresses are present on the failure plane and failure is independent of the frictional characteristics. In the light of this, they concluded that the relatively low compressive strength of the slate concrete may be presumed to be related to the relatively poor particle shape of the aggregate—the compressive strengths of the concretes increased as the particle shape improved and the frictional characteristics changed.

Table 3.4. Properties of some concretes based on slate, marble and granite aggregates (from ref. 15).

	28 day strength (MN/m^2)			*Dynamic elastic properties*		
Aggregate	*Flexural*	*Indirect tensile*	*Compressive*	*Young's modulus (GN/m^2)*	*Shear modulus (GN/m^2)*	*Poisson's ratio*
Slate	5·5	4·8	34·5	42·5	17·7	0·21
Slate	5·3	3·6	29·1	39·9	15·9	0·25
Marble	5·3	3·5	38·2	39·4	16·3	0·20
Granite	5·2	3·2	42·8	38·9	15·7	0·22

Mortar bar tests (ASTM C 227-63 T, ASTM C 227-64) were carried out to determine whether excessive volume changes occurred as a result of alkali–silica reaction. The same high alkali cement used for the concrete specimens was employed at a water/cement ratio of 0·65 by weight. The expansion at 6 months was well below the danger level suggested by ASTM; the tests were continued for 1 year and the linear expansion was still found to be well below the danger level. Very moderate expansion was also found in length change tests (ASTM C 157-60 T, ASTM C 157-64 T) with concrete prisms. The results suggested that the particular slate investigated was harmless in this respect.

Freezing and thawing tests (ASTM C291-61 T) indicated that the structural soundness of the concrete was only slightly affected after 300 cycles and there was no evidence of flecking, popouts or other surface defects.

McGhee and Ozol concluded that, by using the slate aggregate which they investigated, production workable mixes could be made to produce concrete conforming to specifications for two classes of material laid down by the Virginia Department of Highways; and although one of these (column 2, Table 3.3) may be rather harsh as a plain mix, the water-reducing admixture gives excellent workability. They noted that the materials appeared to provide adequate compressive strength, good flexural strength and excellent freeze–thaw resistance and that, furthermore, detrimental volume changes (from the length change tests) seemed to be of little consequence. They pointed out that the slate should perform better, and give higher resistance to flecking in pavement, than some comparable rocks that were already in use in Portland cement concrete. Finally, economic considerations suggested that the use of slate would lead to considerable savings in the particular area in question. And, in fact, slate is currently being used in Virginia for incidental concrete construction as well as for a number of other purposes[16] which will be discussed in Chapter 4.

Another investigation was carried out in connection with the construction of a haul road on the site of the Dinorwic Pumped Storage Scheme in North Wales.[17] The slate was crushed to a grading similar to one given in a United Kingdom specification for cement-bound granular material for sub-base and roadbase construction.[18] The flakey, elongated nature of the slate particles led to the addition of sand to achieve reasonable compaction; and a selected pulverized fuel ash (PFA), of low water demand and of a fineness comparable to cement, was used as a pozzolan which increased the volume of cementitious paste. This gave a good finish and ready compaction of test cubes. The leaner mixes investigated showed sensitivity to water content which could amount to an almost critical change from the too dry to the too wet condition. It was noted that the Vebe test appeared to be more suitable than the slump test for assessing workability with this type of aggregate. Prolonged mixing times did not give rapid loss of workability due to grinding of the slate 10 min mixing producing no significant breakdown. The initial expectation that the shape

Fig. 3.5. Crushed concrete test cube showing the random arrangement of the slate aggregate (from ref. 17, by courtesy of the Central Electricity Generating Board).

of the slate particles may give rise to a laminated arrangement within the concrete did not prove to be correct; as can be seen from the crushed concrete specimen illustrated in Fig. 3.5, the aggregate was found to be randomly orientated. The finish obtained on the test cubes was found to be very good when PFA was added. A summary of the results of this work is given in Table 3.5.

In 1971–2, crushed slate waste was used as dense aggregate in the concrete used for the construction of a gravity dam, over 130 m long, built to form an impounding reservoir of about 47 hectares surface area at Crowdy Marsh in south-west England.[19] The slate waste was screened

Table 3.5. Proportions and properties of concrete mixes using slate aggregate with sand and pulverized fuel ash (from ref. 17).

Batch weights (kg/m^3)					Wet density (kg/m^3)	Slump (mm)	Vebe time (s)	Compressive strength (MN/m^2)		Remarks
OPC	*PFA*	*Slate*	*Sand*	*Water*				*7 days*	*28 days*	
176	250	1 586	167	150	2 300	—	—	3·1	—	Too wet. More sand required.
196	197	1 864 (all in)	—	128	2 369	—	—	13·8	16·7	Poor finish. Sand required.
157	51	1 747	437	88	2 360	—	—	12·0	—	Poor. Difficult to compact. More fines required.
221	133	1 690	266	133	2 300	—	—	17·0	24·5	Good finish. Slight bleeding.
288	73	1 165	582	204	—	80	3	17·3	25·5	Too wet.
273	89	1 092	683	191	2 300	55	4	18·9	30·4	Good finish and workability.
343	85	1 050	656	197	2 350	40	4	24·5	31·4	Good finish and workability.

before being passed through a jaw crusher which produced an all-in aggregate graded from 75 mm down to dust; by normal standards, the aggregate produced was rather flakey and elongated and, with the dust content, looked extremely dirty—furthermore, it was only possible to obtain half the quantity of slate fed into the crusher as aggregate to an acceptable grading. However, the broken stone had a rough surface which ensured a good bond and the dust was an important factor in producing a concrete with a density of 2516 kg/m^3. China clay sand was added to make up a slight deficiency of fines in the slate aggregate. Sulphate-resisting cement was used because of its low heat properties and its ease of supply. It was intended that the concrete should have an aggregate to cement ratio of 8:1, as high strength was not required. However, it was found necessary to use a ratio of 7:1, the cement content of the concrete being 278 kg/m^3; the extra cement, together with a plasticizer, was necessary to obtain workability of the rather harsh mix. In spite of this, the concrete had no slump. Test cubes gave an average 28 day strength of nearly 27 MN/m^2. To obtain a pleasing aesthetic appearance, the slate aggregate on the downstream face of the dam was exposed, using a chemical retarder painted on the shutters before pouring. The Resident Engineer for the project commented that the abrasive nature of the slate particles caused rapid wear in the concrete mixer.

To summarize, there have been instances in which slate has actually been used as the dense aggregate in concrete, both for the manufacture of precast products and for larger scale applications—and there have been a limited number of experimental investigations on the laboratory scale. It has been shown that concrete of about 30 MN/m^2 compressive strength at 28 days can be produced from mixes that have adequate workability characteristics for practical use. Furthermore, McGhee and Ozol's work has suggested that some other properties, including aspects of durability and long-term behaviour, appear to be satisfactory for the particular slate that they investigated.

Generally, the scope for using slate waste is likely to be limited to areas close to its source, particularly where other aggregates are not readily available; under these circumstances, the use of additional cement[15] (and admixtures or other additives) may be justifiable provided that this has no adverse effect on the properties of the concrete.

Finally, it is most important to note that, before any particular source of slate waste is actually used for producing concrete, a proper evaluation must be carried out to ensure that the particular material in question has no detrimental effect upon any aspect of the performance of the concrete under service conditions. The composition of slates from different sources varies considerably and some contain minerals, for example pyrite among others, which can cause problems in concrete.

3.4. EXPANDED SLATE AGGREGATE FOR LIGHTWEIGHT CONCRETE

3.4.1. Lightweight concrete

The normal range of densities for conventional concrete has been given[20] as being between 2200 and 2600 kg/m^3, whereas lightweight concrete usually ranges from about 300 kg/m^3 up to 1850 kg/m^3 or perhaps a little more.

The low density of lightweight concrete, generally achieved by forming internal voids within its structure, has obvious advantages in construction; lighter structures can be built and savings can be made in the foundations as well as in transportation, handling and erection costs. Furthermore, the thermal insulation properties are better and, in this context, it has been argued that some lightweight concretes have the advantages of both load-bearing and thermal insulation properties as well as the added benefit of useful thermal capacity; depending on variations in the inside temperature of a building, the material will either absorb or emit heat, thus providing better control of the internal environment.[21] Lightweight concrete structures have also shown remarkable resistance to shock and blast;[22] in the Second World War, concrete ships withstood the effects of near misses from shells and depth charges—and in Nevada in 1955, during atomic bomb tests on the effects of blast on structures, the only buildings to survive without major damage were constructed from lightweight concrete. Furthermore, many lightweight concretes have good fire resistance because of their lower thermal conductivity and their resistance to spalling.

The properties of lightweight concretes vary enormously over the large range of densities that can be obtained; for instance, concretes with a high volume of voids will tend to have good thermal insulation properties but relatively poor strengths—and, for a low proportion of voids, the reverse tends to apply

There are three main types of lightweight concrete and these can be classified broadly in terms of the different ways in which the voids are introduced. First, aerated concretes can be made by forming gas bubbles in a cement-based slurry which is subsequently hardened to give a solid cellular structure; and materials of this kind will be discussed in Chapter 5. Secondly, lightweight aggregate concretes are made in a similar way to conventional concretes but, in this case, special aggregates with a cellular structure are used. Finally, 'no-fines' concretes contain voids that result from leaving out the fine aggregate. And there are various possible combinations of the three types; for instance, a lightweight aggregate can be used to make no-fines concrete. (There have been a number of relatively brief accounts of lightweight concrete that serve as useful introductions to the subject; for instance, those by Neville,[20] by Spratt[23] and by the Portland Cement Association.[24] For a more detailed account, the reader is referred to the recently published book by Short and Kinniburgh.[25])

The remainder of this chapter is centred upon the use of slate, expanded by heat treatment, as a lightweight aggregate in concrete.

3.4.2. Lightweight aggregates in concrete

The history of lightweight concrete dates back about 2000 years when the Romans used pumice as an aggregate; since that time, other natural materials have also been used, particularly scoria and others of volcanic origin, as well as diatomite. However, most of the progress leading to the artificial lightweight aggregate industry, as it stands today, has been made since the turn of the present century.

By that time, clinker (or cinders in American terminology) was already being used as an aggregate but nowadays its availability in many areas has been affected by the increased use of pulverized coal.[25] Although pulverized fuel ash is now processed into lightweight aggregate in a few countries, the overall production rate is relatively small. Foamed slag was first made in 1911,[25] in Germany, and this material is now being produced and used in a number of countries. Relatively small tonnages of other processed lightweight aggregates, including expanded perlite and exfoliated vermiculite, are also used in concrete.

But probably the most significant group of materials today are the expanded shales, clays and slates (although, as will be seen later, slate provides a very small proportion of the total current production). The First World War gave impetus to the development of this type of material when concrete ships were constructed in the United States as a result of the shortage of steel plate; and this laid the foundations upon which the expanded shale, clay and slate industries eventually grew. However, the lack of technical information, the depression and protection by patents had tended to restrict the initial commercial development in the 1920s and 1930s[22] although, after the Second World War, progress accelerated rapidly.

Nowadays, concretes based on lightweight aggregates tend to be classified into three functional types. However, exact divisions have not been universally agreed, and definitions of the classes tend to vary depending on their source. Nevertheless, the following is a useful guide, but it should not be regarded as anything more than a broad indication since overlap between the three types is often considerable.

Insulating concretes: These are the lightest of the three types, having densities which generally lie between about 300 and 800 kg/m^3 and compressive strengths ranging from below 1 MN/m^2 up to about 4 MN/m^2 or sometimes a little more. They have excellent thermal insulation properties; and according to a recent functional classification recommended by RILEM[26,27] (i.e. Réunion Internationale des Laboratoires d'Essais et de Recherches sur les Matériaux et les Constructions), the coefficient of thermal conductivity of insulating concretes should be below 0·30 W/mK. But their low strength, and often high shrinkage properties, severely restrict their value as structural materials. Exfoliated vermiculite and expanded perlite are commonly used aggregates.

Structural and insulating concretes: These materials represent a compromise between strength and insulating properties for applications, such as blockwork and fill, where both may be of importance. They are stronger and heavier than insulating concretes, ranging up to about 15–20 MN/m^2 and about 1400–1600 kg/m^3. The RILEM classification gives a coefficient of thermal conductivity of below 0·75 W/mK for concretes used for structural and insulating purposes. Pumice and scoria are used for materials in this density range although clinker, or cinders, and a number of other materials that can be used for lightweight structural concrete (see below) are also employed, particularly at the higher end of the density scale.

Structural concretes: These are primarily structural materials that are lighter than conventional concretes but, of course, they do have improved thermal insulation properties too. Compressive strength and density values range from the top of the scale for structural and insulating concrete up to about 70 MN/m^2 and 1900–2000 kg/m^3 respectively. The wide range of strength values has led the CEB (i.e. Comité Euro-International du Béton) to subdivide the structural grades into three shorter strength ranges,[26] namely 20–40 MN/m^2 (with densities of 1600–1750 kg/m^3), 40–50 MN/m^2 (1750–1850 kg/m^3) and 50–70 MN/m^2 and above (1850 1900 kg/m^3). Aggregates suitable for this type of concrete, depending on the properties required, include foamed slag and sintered PFA as well as expanded shale, clay and slate.

A very wide range of aggregates can be used for making lightweight concrete. A recent RILEM classification[26,27] lists a large number of processed and unprocessed materials, obtained both from natural sources and as by-products, as well as some organic materials. The FIP (i.e. Fédération Internationale de la Précontrainte) recently carried out a survey of lightweight aggregate production in a number of countries throughout

Table 3.6. Relative annual production rates, by volume, of different classes of lightweight aggregate (based on data, from the early 1970s in 26 countries, taken from ref. 28).

Processed natural materials	48 %
Unprocessed natural materials	31 %
Processed by-products	10 %
Unprocessed by-products and others	11 %

the world; annual production figures expressed in volume, for various years ranging between 1970 and 1975, have been published[28] for 26 countries. Table 3.6, derived from this data, is an approximate volume breakdown expressed in terms of the types of aggregates produced. (It should be noted that a small proportion of the data has not been included either because it was expressed in terms of weight or because quantities were not given; however, this should make little difference to the overall balance indicated by the table.) Processed natural materials (including exfoliated vermiculite, and expanded perlite, clay, shale and slate) account for about half the total production; and expanded clay is responsible for nearly 90 % of this whereas expanded slate provides less than 2 %. Nearly a third of the total production comes from unprocessed natural materials, most of which is pumice. Processed by-products account for a tenth, over 90 % of which is foamed slag and the remainder sintered PFA. The remaining tenth comprises unprocessed by-products, unspecified materials and others including organics; and nearly 75 % of this is accounted for by clinker, furnace bottom ash and other coal by-products produced in the United Kingdom.

The FIP survey did not include production figures from the United States. Statistics published by the Expanded Shale Clay and Slate Institute in the United States[29] show that lightweight aggregate production in 1947 was about 5·6 million tonnes; and of this, 6·5% was expanded shale, clay and slate, 12·9% was expanded slag and 80·6% other materials. Over the next 20 years, the position changed dramatically. In 1968, total production was about 12·0 million tonnes of which 75·9% was expanded shale, clay and slate, 16·8% expanded slag and 7·3% other materials; of the 75·9% from shale, clay and slate, 67·5% was produced by the rotary kiln process and 8·4% by sintering (and both processes will be discussed in some detail in the next section). Furthermore, by 1971, it was estimated that rotary kiln expanded shale, clay and slate was being used for more than 80% of the structural lightweight concrete placed. Most of the material (generally around 60% by volume) shipped annually by producer-members of the Expanded Shale Clay and Slate Institute, during 1962–77, was used for concrete masonry products;[30] over the same period, much of the remainder was used for structural purposes, and a small but recently growing proportion (up to 10·5% by volume annually) for other uses, including flexible pavement. However, although the majority of producer-members of the Institute operate in the United States and Canada, there are a number in Europe, Asia and South America and the statistics are influenced accordingly.

Other statistics,[31] which give the United States' production of lightweight aggregates in 1972, show that expanded shale, clay and slate provided 63% of the market tonnage, pumice and volcanic cinders 19%, expanded slag, cinders, fly ash aggregate, etc., a total of 15%, and perlite and vermiculite 3%. But these figures do not separate the production of expanded slate from shale and clay. However, another survey[32] pointed out that, in 1970, there were 70 plants in the United States that were producing expanded clay and shale and five plants producing expanded slate—and more than two-thirds of their total production was being used for manufacturing concrete blocks. The survey also quoted United States Bureau of Mines statistics showing that, between 1965 and 1969, the annual production of expanded clay and shale varied between 6·74 and 8·91 million tonnes. Between 1965 and 1968, expanded slate production varied between 0·59 and 0·69 million tonnes per annum. For comparison, over the same period, the annual production of pumice (including pumicite and volcanic cinder) ranged between 2·92 and 3·20 million tonnes and, for expanded perlite and exfoliated vermiculite, the figures were 0·31–0·36 and 0·23–0·26 million tonnes respectively; however, considerable quantities of these latter three materials were being used in applications other than lightweight concrete. The survey also noted that, at that time, only a small proportion of lightweight aggregate was being produced in the United States from the large quantities of fly ash that were available; in 1969, only 0·23 million tonnes were used for this purpose from the 20·2 million tonnes collected.

The present day production of expanded slate aggregate, on a worldwide basis, is difficult to quantify. Nevertheless, there are some figures which give a useful guide. For instance, 330 000 m^3 were produced in Bulgaria in 1972, 55 000 m^3 in Czechoslovakia in 1975, and 60 000 m^3 in the German Democratic Republic in 1975;[28] and, assuming an average bulk density of say 750 kg/m^3, this gives an estimated total annual production rate of about a third of a million tonnes for those countries at that time. And, in comparison, Table 1.4 (p. 7) shows that the quantities shipped or used in the United States in 1972 and 1973 were 1·15 and 0·99 million tonnes respectively. Furthermore, as noted in Chapter 1, expanded slate was also processed in North Wales at about that time although production has now been discontinued. So, although the share of expanded slate in the total processed lightweight aggregate market is only of the order of a few percent, considerable quantities of the material have been produced and used—and the appropriate technologies are already well developed.

Expanded slate concretes tend to lie towards the upper end of the density range (i.e. among the 'structural and insulating' and the 'structural' lightweight concretes) making them comparable, in this respect, to concretes based on processed aggregates such as sintered PFA, foamed slag and expanded clay and shale—and, in fact, the Expanded Shale Clay and Slate Institute suggests that, as far as concrete properties are concerned, there is no distinction between expanded aggregates produced from materials classified as shale, clay or slate.[33]

3.4.3. The development of expanded slate aggregate

The expanded shale, clay and slate industry appears to have originated from the work of S. J. Hayde, who manufactured bricks in the United States at the beginning of the present century.[22] Hayde studied the problem of 'bloating' or expansion that can occur during the firing of bricks and turned it to advantage by developing a process for the manufacture of expanded lightweight aggregate. As mentioned in the previous section, this material was used in the construction of concrete ships in the United States at the end of the First World War—and, at about the same time, Hayde was granted a patent covering his process.

It is not clear when expanded slate was first produced but in 1931 a Russian patent[34] described the use of slate waste, processed by heat treatment, as an aggregate in concrete. And in the early 1930s, Coleman, working at the Building Research Station in the United Kingdom, investigated the expansion of slate in connection with its potential use for making lightweight aggregate.[35] Coleman's work stemmed from the fortuitous discovery that pieces of slate, used as supports during the firing of bricks, bloated as the result of the high temperatures used.[36] Figure 3.6 shows an example of slate expanded by heat; the mechanism of the expansion process, which will be discussed in more detail in the next section,

Fig. 3.6. An example of heat-expanded slate compared with the raw material from which it was produced. (Crown copyright photograph reproduced by courtesy of the Director of the Building Research Establishment.)

depends on the release of gases within the slate—the solid material, softened by the high temperature, is bloated as the result of the internal pressure generated by the gas. Coleman's investigations resulted in a patent[37] and in the publication of one of the first accounts of a systematic study of the properties of expanded slate concretes.[38] Recognizing a potential use for slate waste, he carried out a survey in Wales in 1932 and collected 91 slate samples which he subjected to firing tests to determine their expansion characteristics. But, despite his efforts, there was little interest at the time and the commercial manufacture of expanded slate aggregate in Wales was not started until 1965 and, unfortunately, production ceased a few years later.

By 1940, the use of expanded slate was being considered elsewhere[39] and, at about this time, workers in the United States began to take an interest in the potential of the new aggregate. As the American production figures in the previous section suggest, the commercial development of expanded slate seems to have met with more success there than elsewhere and it is worth considering how this took place.

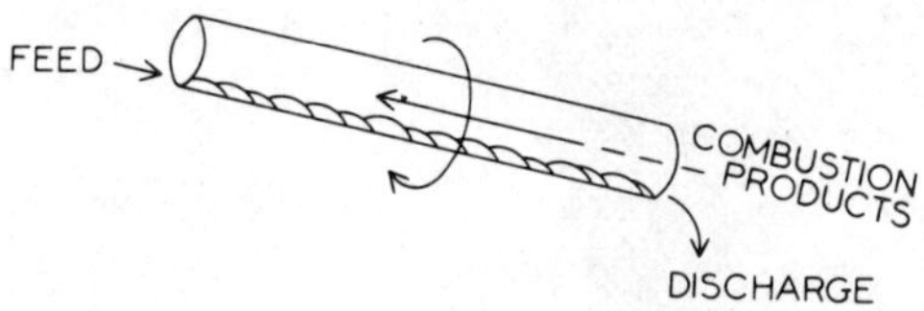

Fig. 3.7. Schematic illustration of the rotary kiln process used for the production of expanded slate.

Although the use of slate was mentioned in one of Hayde's patents, Conley appears to have published the first account, giving experimental results, of investigations carried out in the United States.[40] Conley[41] considered the composition of a number of slates and provided data concerning their expansion properties; he also pointed out that the rotary kiln was already being used for the commercial production of expanded clays and shales and suggested that it would provide a suitable method for expanding slate.

The rotary kiln process is illustrated schematically, in simplified form, in Fig. 3.7; and plant used for the manufacture of expanded slate aggregate by this method is shown in Fig. 3.8. The rotary kiln consists essentially of a long metal tube lined with heat-resistant material. The kiln revolves about its longitudinal axis and is mounted so that one end is slightly higher than the other; the raw material, either crushed to a suitable grading or ground and preformed into pellets, is fed into the higher end and progresses slowly downwards along the kiln as it rotates. The kiln is fired at its lower end using oil, gas or powdered coal and the hot combustion products pass up the kiln in the opposite direction to the feedstock; the feedstock thus encounters a progressively higher temperature as it passes down the kiln and it is then discharged at the lower end. The detailed design and operating conditions of different kilns vary greatly, depending on the material being processed and the required properties of the end product; obviously the rate of rotation and the slope of the kiln determine the time taken for the material

Fig. 5.8. Rotary kiln plant used for making expanded slate aggregate. (Crown copyright photograph reproduced by courtesy of the Director of the Building Research Establishment.)

to pass through it—and the maximum temperature and the temperature gradient are also important factors.

As mentioned in Chapter 1, the Mineral Industries Experiment Station at Pennsylvania State College began a major research programme in 1943[40] which involved investigations into a number of different uses for slate waste, including the production of lightweight aggregate. After exploratory runs on the laboratory scale, the experimental work involved pilot plant trials using an oil-fired kiln, 0·9 m in diameter and 5·5 m long, in order to examine the feasibility of manufacturing expanded slate by means of the rotary kiln process; and this highlighted one of the major problems associated with this method. First, the slate must be raised to a sufficiently high temperature for expansion to occur; but, on the other hand, if the temperature is too high then the individual pieces of slate forming the feedstock tend to adhere to one another and to the inner refractory lining of the kiln. This leads to agglomeration of the slate and to the formation of a ring of material around the inside of the kiln—and, in both cases, the operation of the kiln is restricted. In one instance noted in the report on this work, the difference in temperatures resulting in underfiring and in sticking was as little as 11 °C. The process was also investigated under production conditions, using a full-sized kiln (over 27 m long) and, again, temperature control was found to be critical and confined to a narrow range. Despite these problems, the properties of the experimental aggregates were very promising and compared well with commercially available materials. The report discussed a number of ways of overcoming the agglomeration problem and, like Conley's paper, considered the possibility of pretreating the slate by grinding it and forming the powder into pellets for use as the feedstock.

The publication of the Mineral Industries Experiment Station report, in 1947, came at a time when a great deal of interest was being focused on the production of lightweight aggregates in the United States. Patent protection of Hayde's process had recently expired and this enabled the rotary kiln to be more widely employed on a commercial basis. Furthermore, there was a large potential demand for manufactured lightweight aggregates during the post-war years.

A United States Bureau of Mines Report of Investigations,[42] published in 1948, pointed out that there were serious shortages of these materials; between 1936 and 1946, there had been a more than eightyfold increase in the value of sales by the American building-block industry and there had been an enormous demand for lightweight aggregates suitable for precast building units and for massive structures. Furthermore, the shortage of lightweight aggregates was intensified by the increasing use of powdered coal, with the consequent decline in clinker production. This situation led the Bureau of Mines to undertake a major investigation into the production of lightweight concrete aggregates from clays, shales, slates and a number of other materials, in order to stimulate the commercial manufacture of this type of material. The report described the production of a number of experimental expanded aggregates from crushed slate in a small gas-fired rotary kiln, 0·2 m inside diameter and 3·7 m long, and in a larger oil-fired kiln 0·9 m in diameter and 9·1 m long. (In view of the Mineral Industries Experiment Station report, it is interesting to note that a slate sample from Pennsylvania gave rise to problems with agglomeration and 'ringing' in the small kiln and serious difficulties with temperature control and agglomeration in the larger kiln.)

The technical experience gained from development programmes such as those described above, and the demand for lightweight aggregates, clearly stimulated the commercial development of expanded slate and soon after the war several rotary kiln plants were built.

In 1948, the year in which the Bureau of Mines report was published, production of an expanded slate aggregate under the name 'Solite' was begun by the Southern Lightweight Aggregate Corporation at Bremo Bluff (near Richmond, Virginia). Mocine[43] has given a detailed description of this plant which used two coal-fired rotary kilns (2 m in diameter by 30 m long) giving a production capacity of 150–230 m^3 per day; the raw slate feedstock was quarried and the plant was situated by a railway line which was used for transporting the finished product. The material was crushed to below 9·5 mm, with 10% passing 100 mesh, and had a bulk density of 960 kg/m^3. It gave concretes with densities ranging from 1280 to 1470 kg/m^3 for corresponding cement contents between 220 and 390 kg/m^3, and cement contents of 170 and 420 kg/m^3 gave compressive strengths of 8·6 and 33·1 MN/m^2 respectively. Most of the production was sold to manufacturers of concrete blocks and to ready-mixed concrete producers. By 1954, a further four kilns had been added to the plant, one of which was nearly 2·5 m in diameter and over 33 m long.[44]

In 1953, the Carolina Solite Corporation, a subsidiary of the Southern Lightweight Aggregate Corporation, began the production of Solite from slate near Aquadale, North Carolina. This plant, described by Lenhart,[44] had basic features in common with that at Bremo Bluff; the raw slate was quarried and most of the material produced was transported by means of an adjacent railway line. The plant used two coal-fired rotary kilns, somewhat longer than those at Bremo Bluff, which were lined with Solite concrete for most of their length apart from the hottest zone. The characteristics of the aggregate were the same as that from Bremo Bluff and it was produced in three size ranges: 0–9·5 mm (with a bulk density of 960–1010 kg/m^3) for block manufacture and, for other concretes, a 0–3·2 mm fine aggregate (1090–1110 kg/m^3) as well as a 3·2–19·1 mm coarse aggregate (880 kg/m^3).

Another plant began production in 1953. The product, called Galite, was manufactured by the Georgia Lightweight Aggregate Company at Rockmart, Georgia. Lenhart described the plant[45] which employed two rotary kilns that were nearly 2·5 m in diameter and 38 m long; these were fired with natural gas and had a production capacity of 230 m^3 in 24 hours. Again the plant was served by an adjacent railway line and the raw material was quarried. Three grades of aggregate were produced: 0–4·8 mm (bulk density 830 kg/m^3), 4·8–9·5 mm (620 kg/m^3) and 9·5–19·1 mm (590 kg/m^3). Concrete made from Galite had a density range from 1360 to 1760 kg/m^3, with corresponding compressive strengths from 10·3 to 34·5 MN/m^2; and concrete blocks, with a density of 910 kg/m^3 (cement content about 91 kg/m^3) had a 7 day compressive strength of 7·9 MN/m^2. It is interesting to note that raw slate fines below 4·8 mm, not used as feedstock for the kilns, was sold locally for roadbase stabilization and found a considerable market for this purpose.

The Hudson Valley Lightweight Aggregate Corporation, owned by the Solite Corporation together with local interests, opened a new plant in 1961 at Saugerties, New York. This plant was described by Trauffer[46] and used two rotary kilns, over 3·5 m in diameter and 55 m long, one of which began operating in 1961 and the other in 1962; the total plant capacity was 45 tonnes per hour. As with the other Solite plants, the kilns were coal-fired and the slate was quarried. The plant produced gradings conforming to the ASTM structural lightweight aggregate requirements as well as a 0–9·5 mm material suitable for lightweight asphalt aggregate and another for block manufacture. The products were transported by rail, lorry and barge and about 60% of the total production was carried to the New York metropolitan area by rail.

McCarl[47] pointed out that, by 1964, waste slate was being converted to lightweight aggregate in a rotary kiln (by Pennsylvania Lightweight Aggregate, Inc.) at Bangor, Pennsylvania. Most of the material produced was used for making blocks and was sold in eastern Pennsylvania and in northern New Jersey.

The Vermont Light Aggregate Corporation began production of an expanded slate aggregate (at Castleton, Vermont) late in 1964 and by the beginning of 1967 the production rate was 250 tonnes per day. Trauffer[48] described this plant which was built on a site previously used for the production of roofing granules. Pilot plant trials had been made to investigate the possibility of pelletizing 0–1·2 mm waste slate fines, left from granule manufacture, and using them to produce lightweight aggregate; however, it was established that commercial production was not feasible because of cost and capacity problems. A rotary kiln plant was therefore built to use slate extracted from the existing quarry. The oil-fired kiln was 36 m long and was dumb-bell shaped with a diameter of 2·6 m at the feed end, 2·1 m in the centre section and 3·4 m in the final section. The products were mainly intended for making blocks, and gave compressive strengths averaging 9·7 MN/m^2 from mixes designed for this purpose, but structural grades were also supplied. Bulk densities ranged from about 900 to 1030 kg/m^3. The products were transported by rail and lorry over distances up to 150 miles from the plant.

In all the plants described above, the raw slate feedstock was crushed and sized before processing and the products were similarly treated after expansion. In the Solite plants, the general practice was to provide facilities to air-cool the aggregate after discharge from the kiln—and this was said to temper it by annealing and to make it more inert.[44] But the Galite plant was provided with rotary coolers, one for each kiln, with external water sprays; and air, heated by travelling through the coolers, was passed to the combustion zone to save fuel. Furthermore, the relatively rapid cooling of the product was said to 'freeze' its structure and minimize crystallization and to produce a hard, vitreous material.[45] In the Vermont plant,[48] the product was discharged into a contact cooler and then transferred on to a conveyor equipped with water sprays.

In 1965, the production of an expanded slate aggregate was begun, under the name 'Solite', in the United Kingdom at Wrexham in North Wales. This material was manufactured from slate waste by the rotary kiln process and, by 1972, the production rate was 150 000 tonnes per annum[35] but, as noted in Chapter 1, the plant is now closed because of a pollution problem. Three grades of material were produced: 9·5–19·1 mm (580 kg/m^3), 4·8–9·5 mm (620 kg/m^3) and 0–4·8 mm (830 kg/m^3).[49] Blocks of the material were also sold as a filter medium.[1]

As noted above, the rotary kiln process suffers from difficulties due to agglomeration of the material and ring formation around the inside of the kiln. In 1950, Conley and Ruppert[50] suggested that, because the material must be softened in order to expand, the problem of adhesion is generally inevitable; and although they discussed a number of ways of minimizing it, these were often accompanied by significantly increased production costs. A survey of the literature suggests that, since then, there have been continued efforts to find ways of overcoming these difficulties; generally, the suggested methods that apply to expanded slate production have involved coating the particles in the kiln with a suitable powdered material, such as sand, which does not melt at the temperatures used.

The earliest work seems to have been done in the United States and four American patents were published at about the same time as the commercial production of expanded slate began there.[51–54] The problem was also studied in the United Kingdom and the results of an investigation by the Building Research Station were published in 1960.[55] Various additives were considered and finely ground quartz sand, with a specific surface area of about 200 m^2/kg, was selected. (Lime had been rejected because it promoted the formation of liquid at the surface of the slate particles; and both alumina and magnesia were regarded as being too expensive—furthermore, the presence of magnesia in building products may lead to problems due to the expansion that can occur when it hydrates.) It was found that the addition of 15% by weight of sand to the slate feedstock generally prevented agglomeration and ringing in the kiln, although slight agglomeration was observed when higher temperatures were used. Consideration was given to varying the quantity of sand, and its particle size and grading, as well as feeding it into the discharge end of the kiln and recovering it from the exhaust gases for recirculation. Various relevant patents have appeared since then[56–59] as well as descriptions of other processes involving slate as a raw material.[60,61]

Apart from the problems associated with adhesion inside the rotary kiln, there are other factors which need to be considered. Biege and Cohen[62] pointed out that the rotary kiln process can be thermally inefficient, with about half the heat losses occurring in the exhaust gases and about a fifth in the discharged product; but, by providing the kiln with internal devices to 'tumble' the material and improve the heat transfer, and by cooling the discharged product and returning the heat to the kiln, the fuel costs can be almost halved. Nevertheless, Davies[63] has pointed out that the thermal efficiency of the rotary kiln is low (55% or below), even when it is equipped with elaborate heat recovery arrangements; and he has noted that the sintering process, using a travelling grate, is capable of giving a thermal efficiency up to 85–90% of the theoretical when it is properly controlled. He also noted that abrasion of the slate in the rotary kiln can cause dust in the exhaust gases.

One version of the travelling grate process, suitable for processing slate, involves mixing the crushed and sized raw material with solid fuel and spreading it on to a grate which is in the form of a continuous loop operating in a similar way to a conveyor belt. After the feedstock has been spread on to the grate, it passes under an ignition hood and air is drawn through it, as illustrated in schematic form by Fig. 3.9; the fuel at the top surface is ignited and, as the grate travels towards the discharge end, the combustion zone passes downwards through the bed of material in a similar way to burning tobacco in a pipe. The process is controlled so that the combustion zone has passed completely through the depth of the bed by the time the material is discharged at the end.

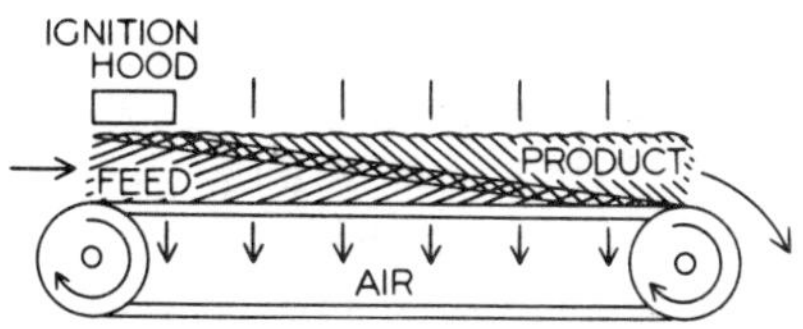

Fig. 3.9. Schematic illustration of the travelling grate process (after Davies[63]).

Apart from the relatively efficient use of fuel, this process has a number of other advantages over the rotary kiln method. Conley and Ruppert[50] have noted that cheap, high-ash coal could be used which would otherwise promote ringing in the rotary kiln; furthermore, raw materials with a low temperature range between expansion and sticking would be more readily processed by this method—and, for the same capacity, the capital cost of the plant would be less.

In 1955, the Exalite Corporation started to produce expanded slate, at North Bangor in Pennsylvania, using a travelling grate furnace with a capacity of 610 m^3 per day; a second unit began production early in 1956 and, by 1958, the production rate was 380 000 m^3 per year.[36] Hoyt[64] pointed out that this plant was built with the relatively low capital cost in mind; the comparatively low fuel consumption of the plant contributed to the low selling price of the product. And the proximity of a good market also helped its commercial success. Crushed slate waste, containing 6% anthracite, was used as the feedstock and three grades of material were produced: 9·5–19·1 mm (bulk density 380 kg/m^3), 3·2–9·5 mm (480 kg/m^3) and 0–3·2 mm (660 kg/m^3). Concrete blocks, made with Exalite and sand, gave compressive strengths of about 8·3 MN/m^2.

A recent report has discussed the possibility of producing expanded slate in North Wales using the travelling grate process.[63]

There is another process which may offer the prospect of producing expanded slate without the problems associated with the rotary kiln method.[65] This is known as the circulation streaming process and involves suddenly subjecting the raw material to high temperature by releasing it, in batches, into a special kiln in which hot gases are circulated; each batch spends a relatively short period in the kiln (for example, 45 s to expand clay pellets) during which individual particles are continually recirculated in a fast-moving stream of hot gas. Ironman[65] has noted that the rapid temperature rise eliminates the premature loss of gases generated within the raw material particles before expansion takes place; and some materials normally regarded as having unsatisfactory expansion characteristics can be successfully processed. He pointed out that aggregate particles with a dense, hard shell can be produced and that the uniform heat treatment which the process provides eliminates the density variation between different particle sizes, which occurs with the rotary kiln method, possibly because of their stratification within the kiln. Ironman described a continuous streaming plant, producing expanded clay, which required lower investment and operating costs than a rotary kiln; the design of the plant incorporated a number of attractive features including automatic control, low heat consumption and considerable operational flexibility regarding shutdown and reheating.

3.4.4. The expansion mechanism and the evaluation of raw materials

Apart from the investigations and developments already mentioned, there have been a number of relatively recent publications reflecting the growing interest in the expansion of materials such as clays, shales and slates in many different countries.[66–70] There have been a number of investigations into the mechanism of the expansion process from the mineralogical and chemical points of view, and some progress has been made with shales and clays.[71] Progress has also been made with slate but fundamental knowledge of the mechanisms involved in its expansion seems to be rather limited.

As indicated in the previous section, the expansion of slate depends on two conditions which are fulfilled at the high temperatures used for the process. First, gases are evolved within the slate particles and the internal pressure which these create causes the material to expand; but, secondly, the material must be in a pyroplastic condition to enable this expansion to take place. Clearly, the ranges of temperature over which these two processes occur can differ considerably, depending on the chemical and mineralogical constitution of the particular material; and it is important that the evolved gases do not diffuse and escape through the pores in the slate before it is sufficiently plastic to expand. Furthermore, it seems likely that the formation of liquid phases, in the pyroplastic condition, will aid the retention of gases and help them to bring about expansion; on the other hand, the softened slate should be sufficiently rigid to support the developing cellular structure—and the slate particles should preferably not adhere to one another or to the inside of the kiln. Ideally the gases should be evolved at the same time that the pyroplastic condition is reached; and there should be a large difference in the temperatures at which incipient fusion and complete fusion take place. From an economic standpoint, the firing temperature should be as low as possible and the retention time in the kiln or furnace should be as short as possible. The finished aggregate particles should preferably have a uniform cellular structure with a vitreous outer skin that minimises water absorption and, of course, should have adequate mechanical properties.

Clearly, it would be desirable to be able to accurately predict the expansion characteristics of a particular slate from a knowledge of its structure and of its mineralogical and chemical composition. Unfortunately, at the present time, this is not possible because of a lack of detailed fundamental information; nevertheless, there is some broad understanding of the mechanisms that are involved in the process.

First, the feedstock particles should be of an optimum size; if they are too big then the temperature increase in their centres may not be high enough—on the other hand, if they are too small, there may not be adequate constraint of the evolved gases to enable satisfactory expansion to occur. It seems likely that factors such as these may be partly responsible for the bulk density variations observed in different size fractions of expanded slate.

The mineralogical and chemical composition of the slate are, of course, fundamental in affecting both its gas evolution and fusion characteristics.

Coleman and Nixon[35] briefly discussed the expansion of slate in terms of mineralogical composition. They suggested that the water evolved from mineral constituents is probably mainly responsible for expansion; they pointed out that chlorite and sericite mica are the most important materials in this respect and they noted that these contain about 12% and 2–6% by weight of water respectively. They also observed wide differences in the cellular structures of expanded materials produced from different slates; these ranged from highly vitreous matrices with large pores to apparently less vitreous materials with relatively fine pore structures.

The Mineral Industries Experiment Station programme[40] included some X-ray diffraction studies on the effect of firing on minerals present in slate. As the temperature was raised, under the conditions employed, the diffraction pattern associated with chlorite disappeared at 700 °C and was replaced by others at higher temperatures, sericite appeared to remain above 900 °C but had disappeared at 1100 °C, and illite had disappeared at

900 °C. Quartz remained at 900 °C but at 1100 °C the lines were weaker, and faint evidence of calcite and dolomite remained at 900 °C but had disappeared at 1000 °C. There have also been other investigations from the mineralogical point of view.[72]

Some attention has been paid to the expansion of slate in the light of its chemical composition. For instance, Conley[41] pointed out that high silica or alumina contents indicate high melting temperatures; he suggested that lower processing temperatures could be achieved by using fluxing agents such as lime or ferrous oxide. Of course, slates contain calcium and iron compounds naturally (see Chapter 1) but the proportions vary considerably and their fluxing action is likely to depend on the mineralogical form which they take. Conley observed that some success had been achieved in interpreting the effect of chemical composition on expansion characteristics by means of triangular diagrams for the SiO_2–Al_2O_3–(CaO, Fe_2O_3, MgO) system; when plotted, materials with similar expansion temperatures were found to fall close together although other factors, such as the alkali contents and the relative quantities of CaO, Fe_2O_3 and MgO, were also involved.

Later studies with clays, by the United States Bureau of Mines,[42] also gave indications of the importance of iron, alkali and alkaline earth compounds but direct relationships between expansion properties and the proportions of these constituents were not very apparent. However, averaged results from a large number of samples suggested that materials which expanded well contained relatively high proportions of alkalies (K_2O, Na_2O), alkaline earths (CaO, MgO) and iron compounds (Fe_2O_3) although there were individual exceptions and inconsistencies. The same investigation included studies with slates and other materials but these gave no clear-cut correlation between expansion characteristics and composition. One conclusion drawn from this work was that more research was needed to fully explain the expansion phenomenon.

The role of iron compounds is evidently complex. Conley[41] pointed out that ferric oxide will react with carbonaceous components that may be present in the slate; this will lead to the formation of expansion gases, such as carbon monoxide, the ferric oxide being reduced to ferrous oxide which can then act as a fluxing agent. (There have also been a number of other publications which have referred to iron oxide in the context of expanding slate and other materials.[73–75])

Pyrite and calcite are found in some slates and their presence is likely to give rise to the evolution of gases during the expansion process. For instance, it may be possible that pyrite reacts with oxygen to form sulphur dioxide, as a gas, together with ferrous oxide.[66] And it has been suggested that the expansion of certain clays and shales results from the decomposition of calcite and the consequent evolution of carbon dioxide;[71] but, on the other hand, the presence of calcite may be undesirable since it can aggravate the problems associated with sticking in the rotary kiln process.[63] Dolomite is another carbonate which has been found in slates and which may possibly provide carbon dioxide for expansion.

Clearly the detailed processes involved in fusion and gas evolution are very complex and are dependent on the detailed chemical and mineralogical composition of a particular slate. In the absence of adequate understanding of the large number of variables involved, and their interactions, the assessment of a potential raw material for manufacturing lightweight aggregate must be based on empirical methods. A number of publications already mentioned have suggested how such an assessment might be approached but the United States Bureau of Mines have published an information circular,[76] by Hamlin and Templin, which describes a comprehensive, detailed procedure for evaluating clays, shales and slates as raw materials for rotary kiln production. Despite the reference to this specific process, the approach and preliminary laboratory tests described seem to be of more general interest and perhaps wider applicability. (It is interesting to note that Hamlin and Templin suggest that many problems encountered in using the rotary kiln method have resulted from inadequate initial testing of the raw material, to determine the proper processing techniques, and the failure to take proper account of the processing characteristics of the raw material in designing and specifying the equipment used.)

Hamlin and Templin's report describes a three-stage procedure for sampling the deposit of the material to be evaluated. If initial samples show promising expansion characteristics, then systematic sampling of the deposit is used to establish its boundaries and the quantity and uniformity of the material which it contains. Finally, after evaluation of these samples, large samples are taken for testing in a rotary kiln.

The laboratory evaluation begins with drying tests, both in the air and in an oven at 220 °F (104 °C), to determine whether the material disintegrates or spalls under these conditions; this is important since fines produced by spalling in the drying and preheating zones of the rotary kiln will tend to cause agglomeration in the bloating zone. (Furthermore, it can also cause high levels of dust in the exhaust gases from the kiln.[63])

The oven dried material is used to determine the crushing characteristics under prescribed conditions; these characteristics are most important, particularly in the case of shales and slates, again because the presence of more than 10% fines (below 8 mesh) will tend to cause sticking in the kiln. Apart from a sieve analysis, the crushed particles are examined for shape, thickness, type of fracture and the presence of impurities. Hamlin and Templin pointed out that there are particular problems associated with slate because its particles tend to split along the cleavage direction and can pass through a jaw or roll crusher as thin plates; such particles are difficult to size properly and their thin cross-section does not allow good expansion. They suggested that a toothed roll crusher will break the slate particles across the cleavage planes and give a thicker cross-section.

They recommended small-scale firing tests of two kinds, slow and rapid. The slow firing tests, which are not essential, involve the use of briquets moulded from material ground to below 20 mesh and then mixed with water to form a plastic mass. After drying, the briquets are placed in an electric muffle furnace and the temperature is raised relatively slowly, by 1200 °F (667 °C) in 3 hours, and then more rapidly up to 2400 °F (1316 °C); briquets are withdrawn at different temperatures between 1800 °F (982 °C) and 2400 °F (1316 °C) and properties such as colour, hardness, specific gravity, water absorption and shrinkage or expansion are determined. These tests indicate the expansion range as well as sticking and melting temperatures; they are valuable in indicating materials with a short expansion range which are identified by melting or glazing.

The quick firing test is the most important preliminary test and gives a fairly accurate indication of the actual processing conditions required. Twenty-gram samples of 6·4–12·7 mm sized material, in fire-clay boats, are placed in a preheated kiln for fixed periods; the firing periods, or retention times, used are 5, 10 and 15 min and the temperatures range, in 100 °F (56 °C) steps, from 1800 °F (982 °C) up to 2400 °F (1316 °C), or until overfiring and melting occurs.

Figure 3.10 shows the effect of firing temperature on the density of a slate aggregate after retention times of 5, 10 and 15 min; clearly, at least 10 min is required for proper expansion under the conditions employed. Corresponding curves for a shale indicated that a similar time was required but the results for a clay showed little variation for the three retention times used. Figure 3.11 compares the results for all three materials with a retention time of 15 min; the more refractory nature of the slate is evident, although when a sufficiently high temperature is reached its expanded density is as low as that of the other materials, if not lower. Hamlin and Templin pointed out that curves obtained for a given material with different sized particles will have the same general characteristics; but, with larger

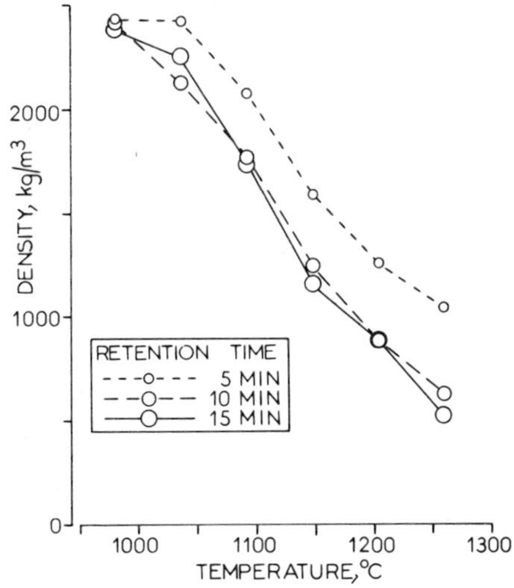

Fig. 3.10. The effect of firing temperature on the density of a slate (after Hamlin and Templin[76]).

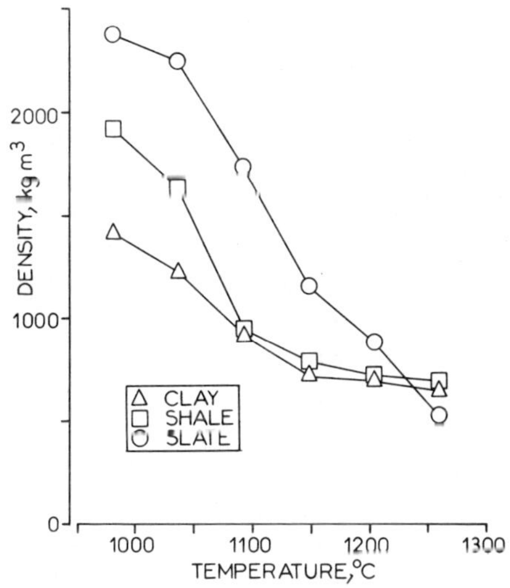

Fig. 3.11. A comparison of the effect of firing temperature, with a retention time of 15 min, on the density of a clay, a shale and a slate (after Hamlin and Templin[76]).

particles, higher temperatures will be needed for the same expansion using a constant retention time. In establishing conditions for rotary kiln tests, they recommended determining individual curves for the component size fractions of the feedstock and combining them. Furthermore, variations in the composition of the raw material can often be detected by differences in the density/temperature curves; thus such tests are useful in determining the uniformity of a deposit.

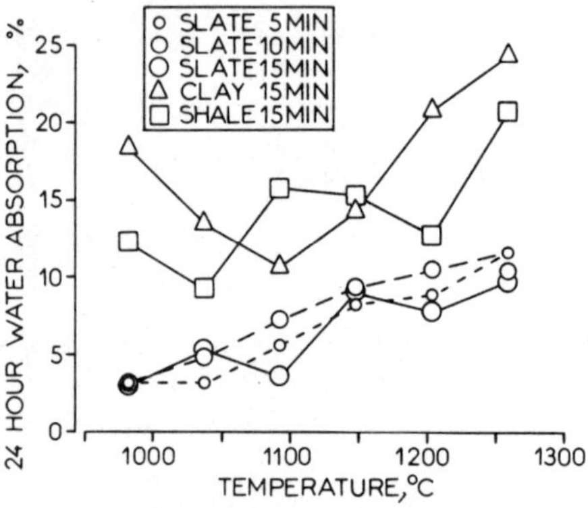

Fig. 3.12. The effect of firing temperature on the water absorption, by weight, of a clay, a shale and a slate (after Hamlin and Templin[76]*).*

As will be seen in the following section, the water absorption of a lightweight aggregate is of great practical significance in its use for making concrete; it is desirable that the water absorption value should be as low as possible. Figure 3.12 shows 24 hour water absorption data corresponding to the variables in Figs. 3.10 and 3.11; although the results appear to be rather scattered (and the sensitive determination of water absorption can be extremely difficult with lightweight aggregates), the expanded slate shows relatively low values—and where the densities of the materials are similar (i.e. after 15 min at 2300 °F (1260 °C)), the water absorption of the slate appears to be less than half that of the other materials.

Hamlin and Templin commented on methods for evaluating various properties of the experimental aggregates including specific gravity, water absorption, particle structure and a qualitative indication of strength. They also described a procedure for evaluating the aggregates' performance in concrete, using atmospheric or high-pressure steam curing, and for assessing their pozzolanic properties. Finally, they described detailed procedures for testing the raw materials in a pilot-scale rotary kiln; such tests, carried out in the light of the preliminary laboratory evaluations, are useful for confirming or adjusting process parameters such as retention time, temperature conditions and sizing of the raw material as well as indicating peculiarities in the processing characteristics that might affect production on a commercial scale. The kiln they used was 4·3 m long with an internal diameter of 0·4 m; the slope, rate of rotation and feed rate could all be varied over a wide range and the position of the burner, which used gas, was adjustable.

The procedures described by Hamlin and Templin provide a good basis for a rapid, preliminary evaluation of possible raw materials for lightweight slate aggregate production; apart from sampling, a complete appraisal can be made in 3 days and, although the approach is directed towards the rotary kiln process, the laboratory tests seem to be of more general applicability and of course can be modified or extended as necessary. Obviously, before commercial production is contemplated, full-scale tests are necessary and a complete evaluation of the properties of concretes based on the expanded slate must be carried out; furthermore, it is necessary to ensure that the material complies with the requirements of any standards that apply in the market areas for which it is produced, and in connection with the applications for which it is to be used.

3.5. THE PROPERTIES OF EXPANDED SLATE CONCRETES

3.5.1. Mix variables

The properties of a lightweight aggregate concrete depend very much on the characteristics of the particular aggregate used; and even aggregates of the same generic type vary considerably depending on their source and on the processing conditions used in their manufacture. Thus, different expanded slate aggregates can give fairly wide variations in the properties of concretes in which they are used, and so the characteristics of any particular material should be properly understood before it is employed. It follows that responsible engineering practice involves adherence to the manufacturer's recommendations in the use of a specific material and, furthermore, the commercial development of an experimental aggregate must require proper evaluation of its performance in concrete both in the fresh and hardened condition.

Of course, the properties of lightweight aggregate concrete also depend very much on mix design. Detailed and specific examination of this subject is beyond the scope of the present book and there are numerous discussions elsewhere, some of which are listed at the end of this chapter;[77–84] nevertheless, a mix design chart, reproduced in Fig. 3.13, is included here because it refers specifically to an expanded slate aggregate, namely the material that was manufactured some years ago in the United Kingdom. (In fact, Fig. 3.13 illustrates one of a number of similar mix design charts, shown by Spratt,[83] for a variety of lightweight aggregates. These charts come from the work of Owens,[85] who based this straightforward approach to lightweight aggregate concrete mix design on maintaining constant proportions, both of the water and of the coarse size fraction of the particular aggregate under consideration, in relation to the volume of the compacted concrete; thus the variables are the cement content and the proportion of fines and, in some cases, varying quantities of air-entraining agent are used.)

It will, of course, be necessary to consider the general effect of mix variables on the properties of expanded slate concrete and, as background to this discussion, it may be useful to briefly consider some of the special characteristics of lightweight aggregates in this context.

The workability of the fresh concrete and the properties of the hardened material depend on factors including the cement content and the aggregate/cement ratio, the type and grading of the aggregate, the replacement of lightweight fines by sand, the proportion of water used and the employment of admixtures. As might be expected, higher cement contents and lower water/cement ratios generally give concretes with higher strength and elastic modulus values, and other improved performance characteristics. But one difficulty with lightweight aggregates lies in the fact that their water absorption is often very high and that the rate of absorption for different aggregates varies considerably. Moreover, the absorption characteristics of the different size fractions of the same aggregate very often vary as well. This frequently tends to preclude using water/cement ratio as the basis for mix design because of the uncertainty of how much water has been absorbed by the aggregate at any given time after mixing; furthermore, continued absorption beyond the mixing period can create problems with workability. Workability can also be adversely affected by the particle shape and texture of many types of lightweight aggregate but it can generally be improved by using a relatively high proportion of fines and also by partial replacement of the fines with natural sand. Sand replacement may lead to a reduction in the water requirement and sometimes in the cement content too; and it can have a marked effect on the properties of the

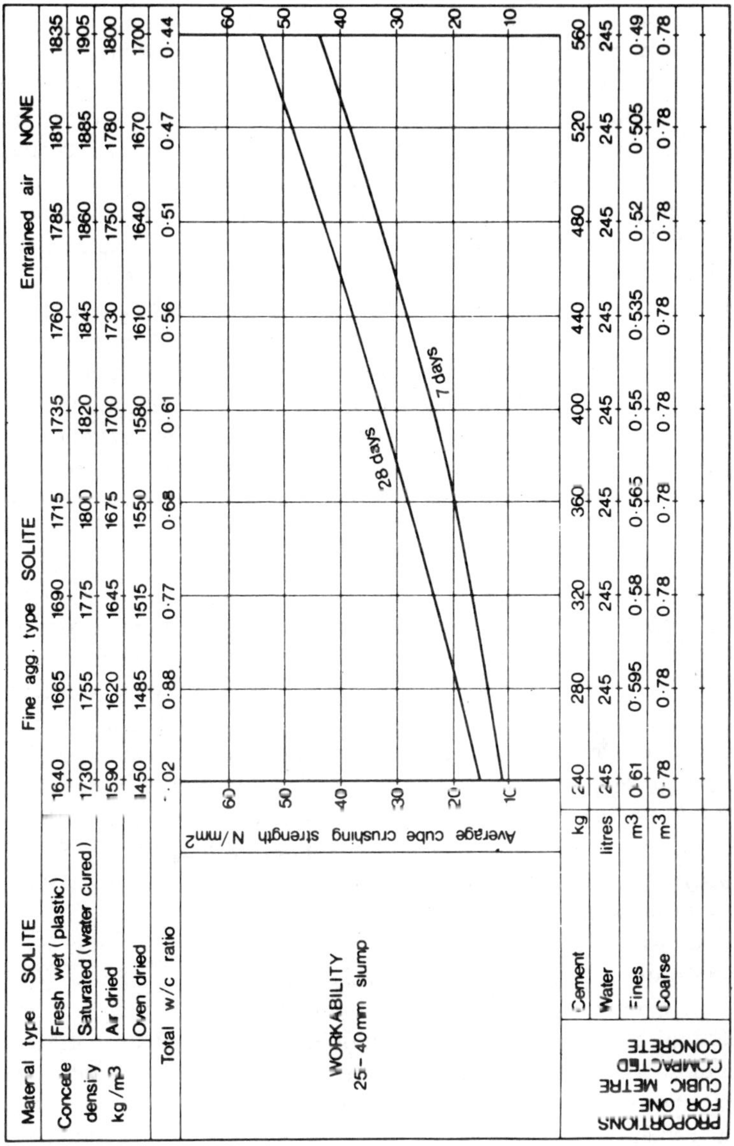

Fig. 3.13. *Mix design chart for lightweight aggregate concrete based on an expanded slate aggregate that was, at one time, manufactured in the United Kingdom (from ref. 83, reproduced by courtesy of the Cement and Concrete Association).*

hardened concrete. For instance, the density is generally increased, and the thermal insulation properties reduced, but compressive strength and other mechanical properties are sometimes improved; furthermore, drying shrinkage may be lowered. Air entrainment can be used to promote workability and to reduce bleeding and segregation as well as the water requirement, and it has been found to be particularly useful with some harsh mixes; it often tends to significantly reduce the density of lightweight concrete still further and there have been circumstances noted under which it has appeared to improve the strength up to a certain air content.[86]

The interactions between the mix variables are extremely complex and it is sometimes difficult to disentangle direct 'cause-and-effect' relationships. Nevertheless, since the mid-1930s, there have been a number of research programmes involving investigations into concretes based on expanded slate aggregates, and these have led to a broad picture of the range of property values that can be achieved. Some of the more important work that has been published is reviewed below; after a brief survey, indicating the scope of the major publications and putting the work into historical perspective, the properties of expanded slate concretes will be considered in some detail.

3.5.2. Historical background

As noted in Section 3.4.3, Coleman, working in the United Kingdom, published one of the earliest accounts in 1936.[38] He examined the properties of various relatively low strength concretes in the form of precast partition slabs; he used expanded slate aggregate with aggregate/cement ratios of 6 and 11 by volume and compared these materials with pumice, coke breeze and clinker concretes using mixes that were no richer than 6. He also considered the possibility of using expanded slate aggregate in reinforced concrete and tabulated suitable mixes, with aggregate/cement ratios ranging from about 2·6 to 6·0 by volume, using all-slate aggregate as well as sand replacement.

There seems to have been relatively little activity during the years immediately following the publication of Coleman's paper but shortly after the Second World War, several relevant publications appeared in the United States, including two already mentioned in Section 3.4.3. First, the Mineral Industries Experiment Station report[40] quoted data for concretes using experimental expanded slates, with no sand replacement, and aggregate/cement ratios ranging from 4 to 8 by volume. Secondly, the Bureau of Mines Report of Investigations[42] gave the results of tests, by the National Bureau of Standards, on concretes made from experimental materials with no sand replacement and using aggregate/cement ratios varying between 2·85 and 9·0 by volume.

In 1949, Kluge, Sparks and Tuma reported the results of studies on lightweight aggregate concretes that were carried out by the United States National Bureau of Standards.[86] This work was prompted by the introduction of the new lightweight aggregates that were being used as a result of pressure from shortages of housing and materials, and because of their thermal insulation properties; technical information available up to that time was sparse, and tended to be limited to cinders and expanded shales and slags, so the National Bureau of Standards programme involved the detailed comparison of concretes based on eleven lightweight aggregates and on sand and gravel. The lightweight aggregates included nine commercial materials (pumice, exfoliated vermiculite, sintered fly ash and diatomite, expanded perlite and shale, and three expanded slags) as well as two experimental expanded materials made from slate and clay. The results for one of the expanded slates quoted in the Bureau of Mines report appear to be the same as those given by Kluge, Sparks and Tuma for their expanded slate concrete; but in addition to these results, they provided data from a larger range of tests and compared them with those obtained from the wide variety of other concretes which they examined. An interesting aspect of their report was the attention they paid to the special problems arising from the character of the lightweight aggregates, particularly in connection with proportioning and mixing. They noted that, for reasonable workability, a much greater proportion of fine material (passing a United States No. 4 sieve) was generally needed compared with sand and gravel aggregates and they also used an air-entraining agent. Water absorption by the aggregates caused them some problems; they pointed out that differences in absorption rates and in the total values of different sized aggregate particles precluded using water/cement ratio as a basis for mix design. Mixes were therefore proportioned by volume, using aggregate/cement ratios between about 2 and 10 without sand replacement, and the water requirement was based on slump tests. They noted that slump values of 102–127 mm, with air entrainment, generally gave mixes that could be cast satisfactorily, but the overall range that they used varied between 76 and 178 mm.

In 1957, the results of another comparative study in the United States were published by Shideler.[87] This work was prompted by increasing interest in structural applications for lightweight aggregate concretes and Shideler compared sand and gravel concretes with those using expanded aggregates, made from slate, clay, slag and several shales, without sand replacement. He based his mix designs on cement content, using the minimum proportion of fine aggregate (i.e. passing a No. 4 sieve) which allowed satisfactory workability, and used a slump value of 64 mm to gauge the water requirement. Apart from some trial mixes, the major part of the work involving expanded slate concretes employed two mixes with aggregate/cement ratios of 2·2 and 3·0 by weight, both with air entrainment.

The manufacture of expanded slate in the United Kingdom, which began in 1965, was followed by a number of publications from Swamy and his colleagues.[88–94] They investigated a wide range of properties of concretes based on this material and quoted aggregate/cement ratios ranging from between 2·0 and 4·1 by weight—and they sometimes used sand replacement. There were also a number of publications which described concrete structures in which the British expanded aggregate had been used.[95–97]

3.5.3. Compressive strength

Compressive strength is one of the most generally quoted properties in the field of concrete technology; compression tests are relatively simple to carry out and are widely used in research and development, quality control and for specification. Furthermore, compressive strength can often be related empirically to other properties.

Coleman's paper[38] provides a considerable amount of information relating to compressive strength properties. His low cement content concretes, with aggregate/cement ratios of 6 and 11 by volume, had air dry strengths ranging from 1·7 to 7·6 MN/m^2 (at ages of at least 28 days) with air dry densities from 560 to 1280 kg/m^3; and, as might be expected, the lower of these ratios tended to give the stronger and denser materials. The strength values generally compared well with the concretes based on pumice, breeze and clinker, and in some cases were considerably higher. Coleman also prepared a series of all-slate concretes, using equal weights of 0–4·8 and 4·8–9·5 mm size fractions, in which he varied the aggregate/cement ratio from 2·6 to 4·4 by volume (1·2 to 2·1 by weight) with corresponding water/cement ratios from 0·625 to 0·92. Some of Coleman's results, together with others, are plotted in Fig. 3.14 and it can be seen that, in progressing along this series with increasing aggregate content, the strength and density fall—and Coleman noted that, at the same time, the mixes tended to become harsher and more prone to segregation. Nevertheless, his two best mixes gave good, workable concretes with 7 day compressive strengths of 18·8 and 17·9 MN/m^2 and corresponding

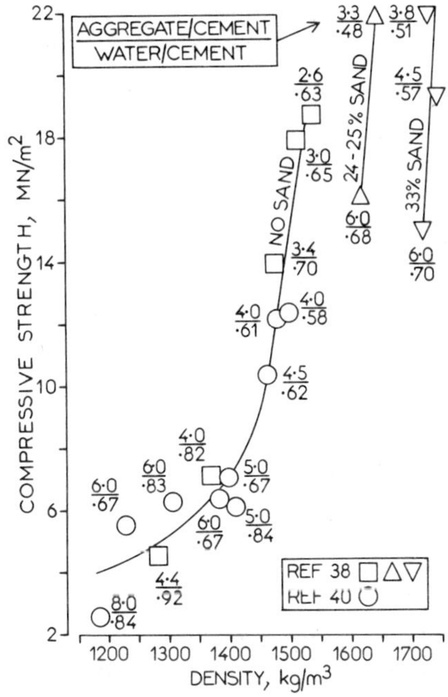

Fig. 3.14. The relationship between 7 day compressive strength and density for some expanded slate concretes, showing the effect of aggregate/cement ratio (by volume), water/cement ratio and sand replacement (by volume). (Data from refs. 38 and 40.)

densities of 1540 and 1510 kg/m^3; their respective aggregate/cement ratios were 2·6 and 3·0 by volume (1·2 and 1·4 by weight) and the water/cement ratios were 0·625 and 0·65.

Coleman also prepared various concretes using sand replacement and found 7 day compressive strengths ranging from 15·0 to 22·0 MN/m^2 and densities from 1620 to 1740 kg/m^3. He provided data for five mixes in which he used two different sands (0–3·2 mm and 0–4·8 mm replacing 24–25 % and 33 % respectively of the total volume of the aggregate, corresponding to 53 and 65 % by weight). He also altered the water/cement ratio and aggregate/cement ratio as well as the slate grading. Coleman's results show that, at each sand replacement level, increased aggregate/cement ratio and water/cement ratio led to significantly harsher mixes; and, in both cases, Fig. 3.14 shows a corresponding decrease in strength with little change in density. The higher replacement level, with the coarser sand, gave markedly higher densities but a similar strength range to the lower level over a roughly

Table 3.7. Mix composition and some properties of concretes made from experimental expanded slate aggregate (from. ref. 40).

Cement/aggregate ratios by volume (cement/fine/coarse)	*Water/cement ratio*	*Slump (mm)*	*Compressive strength (MN/m^2)*		*Density (kg/m^3)*	
			7 days	*28 days*	*7 days*	*28 days*
1:2:2	0·58	25	12·34	13·79	1 499	1 346
1:2:2	0·61	64	12·14	17·93	1 479	1 368
1:2½:2	0·62	nil	10·34	14·48	1 462	1 442
1:2:3	0·67	25	7·03	11·86	1 398	1 346
1:2½:2½	0·84	nil	6·10	11·55	1 410	1 362
1:3:3	0·67	nil	6·36	8·55	1 382	1 378
1:2½:3½	0·67	nil	5·49	6·21	1 229	1 201
1:2½:3½	0·83	25	6·24	7·93	1 307	1 314
1:4:4	0·84	nil	2·57	6·76	1 185	1 185

N.B. 9% water allowed for absorption.

comparable range of aggregate/cement and water/cement ratios—and the mid-point of this strength range roughly corresponds to the strongest, but significantly lighter, all-slate concretes although these had relatively low aggregate/cement ratios.

The work carried out by the Mineral Industries Experiment Station[40] involved all-slate mixes with no sand replacement. A preliminary trial, using slate expanded on the laboratory scale, had given a concrete with a density of 1540 kg/m^3 and a compressive strength of 13·9 MN/m^2 at 16 days, and Table 3.7 shows details of some concretes made with expanded slate from subsequent pilot plant trials. The mix proportions are expressed in terms of rodded volumes, and the division between fine and coarse aggregate fractions is roughly represented by the 8 mesh sieve size—95% of the coarse fraction passed 32 mm and 75% passed 19 mm. Nine percent water was allowed for absorption by the aggregate. As Table 3.7 indicates, changing the total aggregate/cement ratio from 4 to 8 (and varying the water/cement ratio between 0·58 and 0·84) resulted in a general reduction in density and strength, and the 7 day values are compared with Coleman's results in Fig. 3.14.

Table 3.8 is derived from the United States Bureau of Mines report[42] and shows the compressive strengths of a number of concretes, using high early strength cement, made from experimental expanded slate aggregates and tested by the National Bureau of Standards. Aggregates A and B were

Table 3.8. Mix composition and some properties of concretes made from experimental expanded slate aggregate (from ref. 42).

	Aggregate						
	A	B		C			
		Mix 1	Mix 2	Mix 1	Mix 2	Mix 3	Mix 4
Mix proportions (by volume):							
Cement	1·00	1·00	1·00	1·00	1·00	1·00	1·00
Aggregate							
fine fraction	2·50	1·20	—	4·05	2·34	1·67	1·28
coarse fraction	2·50	4·55	6·70	4·95	2·36	2·03	1·57
Bulk density of aggregate (kg/m^3):							
fine fraction	801	1 009	—	1 169	1 169	1 169	1 169
coarse fraction	705	753	753	625	625	625	625
Cement content (kg/m^3)	312	301	217	167[a]	279[a]	390[a]	502[a]
Density of fresh concrete (kg/m^3)	1 346	1 506	—	1 233	1 297	1 394	1 490
Compressive strength (MN/m^2):							
7 days	19·03	18·93	11·55	5·23	11·58	20·00	25·99
28 days	—	—	—	8·96	15·86	23·79	27·58
Strength/density ratio:							
7 days	0·014	0·013	—	0·004	0·009	0·014	0·017
28 days	—	—	—	0·007	0·012	0·017	0·019
Workability	Good	Fair	Very poor	Fair to good	Good	Very good	Very good

[a] Approximately.

obtained from the same source of slate waste but processed under different conditions; and aggregate C was obtained from the waste tips of an abandoned quarry close to the source of A and B.

As mentioned earlier, Kluge, Sparks and Tuma[86] made a detailed comparison between eleven lightweight aggregates without sand replacement. Each of these, and a sand and gravel aggregate, was used to prepare four concrete mixes employing high early strength cement with a range of cement contents from about 170 to 500 kg/m^3. For the slate aggregate, the No. 8 sieve very roughly represents the division between the fractions described as fine and coarse in so far as 90 % of the fine material and 11 % of the coarse passed this size—and 99 % of the coarse material passed 9·5 mm with 55 % passing No. 4. In general, over the range of cement contents used, the highest 7 and 28 day compressive strengths were obtained with the expanded shale, clay and slate aggregates (and both the 7 and 28 day results quoted for the slate are identical to those for aggregate C in Table 3.8). The shale concretes were markedly stronger than the others at higher cement contents and the clay concretes were somewhat stronger than the slate—and this order was reflected in crushing strength values obtained from the three aggregates—but at cement contents of about 170–180 kg/m^3 they gave very similar strengths. For the lightweight concretes having 28 day compressive strengths in the 16–28 MN/m^2 range, the expanded slate had the lowest room dry densities (1270–1460 kg/m^3).

Shideler's work,[87] also involving the comparison of a number of lightweight aggregates, required preliminary mixes to determine the proportions needed to produce concretes having nominal 28 day compressive strengths of 21 and 31 MN/m^2. The aggregates were screened and recombined so that the grading of the fines corresponded to the appropriate specification (ASTM C 330-53 T) for lightweight aggregates, with 85 % passing a No. 8 sieve; the coarse material contained equal weights of the No. 4–9·5 mm and 9·5–19·0 mm fractions. The trial mixes, using the slate aggregate both with and without air entrainment, gave strengths ranging between about 16 and 31 MN/m^2 with fresh concrete densities from 1570 to 1700 kg/m^3; these corresponded to cement contents from about 280 to 450 kg/m^3 and water/cement ratios from about 1·02 to 0·58 (with air dry aggregate). The nominal strength mixes, with air entrainment, used water/cement ratios and cement/fine/coarse aggregate ratios of 0·76 and 1·00:1·96:1·08 (by weight) for the 21 MN/m^2 material, and 0·59 and 1·00:1·32:0·88 for the 31 MN/m^2 material. The fresh concrete densities were 1560 and 1640 kg/m^3 respectively with air contents of 7·7 and 6·8 % by volume—and the actual 28 day wet compressive strengths were 19·4 and

30·0 MN/m^2. The cement contents required for the expanded slate in the nominal strength mixes were close to the average values for all the lightweight aggregates used. Shideler also provided data showing the development of compressive strength with time and this is plotted for the expanded slate concretes in Fig. 3.15. The figure shows that, for continuous moist curing, the strength increase between 28 days and 1 year was 40 % and 21 % for the 21 and 31 MN/m^2 materials respectively; with moist curing for

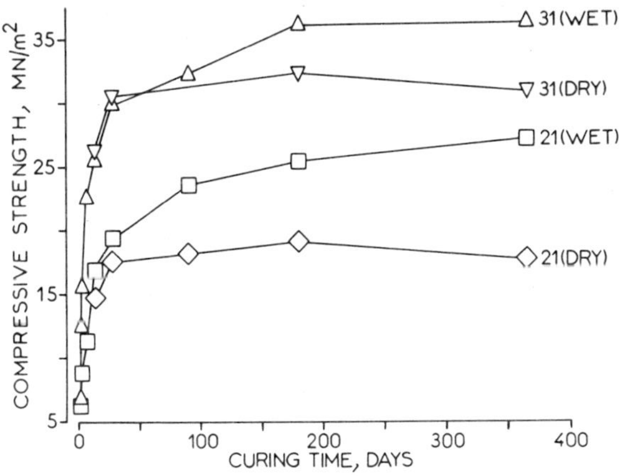

Fig. 3.15. The development of compressive strength for expanded slate concretes with nominal compressive strengths of 21 and 31 MN/m^2, under wet (continuous moist curing) and dry (50 % relative humidity after 7 days' moist curing) conditions (after Shideler[87]).

7 days and subsequent storage at 50 % relative humidity, the compressive strength decreased slightly after about 3–6 months—and this was attributed to shrinkage stresses.

Swamy and his colleagues have also provided some useful compressive strength data. For instance, Bandyopadhyay and Swamy[90] prepared a series of expanded slate concretes using aggregate/cement ratios which varied, in four steps, from 4·10 down to 2·22 by weight with cement contents increasing from 230 to 400 kg/m^3, the corresponding water/cement ratios required to achieve full compaction ranged from 0·82 to 0·45. The division between the fine and the coarse aggregate which they used was represented by the 4·8 mm sieve size; as the aggregate/cement ratio was reduced, the fine/coarse aggregate ratio was lowered from 1·20 to 1·04 by weight. For each of the four aggregate/cement levels, mixes were prepared in which all-slate aggregate was used and others in which two-thirds the weight of the lightweight fines was replaced by natural sand. As Table 3.9 shows, the sand replacement resulted in a reduction in the water/cement ratio that varied between 4 and 6 %, and the dry density at 28 days was increased by between 5 and 7 %; but the effect on the compressive strength was relatively small and even gave a reduction in one case. Taking the results as a whole, the 28 day compressive strength increased considerably as the aggregate/cement

Table 3.9. The effect of partial sand replacement (two-thirds of the fines by weight) on water/cement ratio to achieve full compaction, and compressive strength and dry density at 28 days (from ref. 90).

Cement/aggregate ratios by weight (cement/fine/coarse)	*Sand used*	*Water/cement ratio*	*Compressive strength (MN/m^2)*	*Dry density (kg/m^3)*
1·0:2·24:1·86	no	0·82	23	1 620
	yes	0·78	24	1 710
1·0:1·91:1·62	no	0·73	31	1 660
	yes	0·70	32	1 780
1·0:1·27:1·18	no	0·53	44	1 700
	yes	0·50	42	1 800
1·0:1·13:1·09	no	0·47	52	1 750
	yes	0·45	54	1 840

ratio was reduced from 4·10 to 2·22, but the corresponding density increase was slight.

Data obtained from the sources discussed above, and others, are plotted together in Fig. 3.16 to give a general view of the relationship between compressive strength and density for expanded slate concretes; in most cases, 28 day strength values have been used (and, in one instance, 'at least' 28 days) but a few sources have quoted typical values without reference to age. Where available, individual values are shown but otherwise ranges are indicated—and the strength/density relationship obtained from the mix design chart in Fig. 3.13 is also included. It is not clear from some of the sources cited whether sand replacement has been used, so no distinction between sanded and non-sanded mixes is made in the figure. Furthermore, it should be noted that some concretes were prepared with high early strength cement, and these results are distinguished from the rest. Swamy and Ibrahim[88] used ultra fine Portland cement, that also developed high early strength, to produce expanded slate concretes with compressive strengths of about 30–40 MN/m^2 at 1 day and 55–60 MN/m^2 at 28 days; for these materials, they employed an aggregate/cement ratio of 2·00 by weight

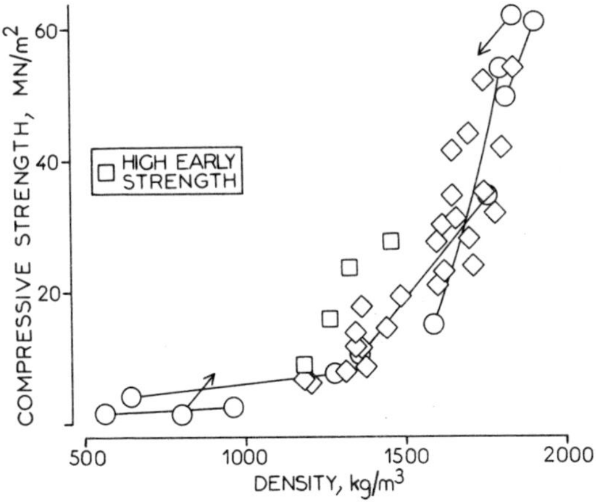

Fig. 3.16. The relationship between compressive strength and density for a range of expanded slate concretes. (*Data from refs.* 23, 38, 40, 45, 83, 86, 87, 90, 95, 98 *and* 99.)

and a water/cement ratio of 0·44. In these mixes, the coarse aggregate was graded from 9 to 19 mm and the fine aggregate, amounting to 44% of the total aggregate weight, consisted of varying proportions of fines and sand.

3.5.4. Modulus of rupture

Coleman[38] found that, for the relatively low density, low strength, precast partition slab materials which he examined, those based on expanded slate aggregate gave a spread of modulus of rupture values of 0·7–1·4 MN/m^2 when tested wet after curing for at least 28 days. This range was comparable with, but extended higher than, that of the pumice concretes (and above the British Standard requirement for precast concrete partition slabs)—and it fell within the range for the breeze and clinker concretes, although these tended to have slightly higher densities.

Kluge, Sparks and Tuma[86] plotted compressive strength against modulus of rupture at 28 days for a number of the lightweight concretes which they investigated; and this plot suggested that, in comparison with the other materials, the expanded slate concretes had relatively high modulus of rupture values (2·0–3·4 MN/m^2) for their range of compressive strengths (9·0–27·6 MN/m^2).

Shideler[87] compared the modulus of rupture values of his two expanded slate mixes (having nominal compressive strengths of 21 and 31 MN/m^2) with corresponding mixes based on six other lightweight aggregates and on a sand and gravel aggregate. For the 21 MN/m^2 series, the expanded slate gave the highest modulus of rupture after 28 days' moist curing (3·5 MN/m^2) but after 90 days the value (3·7 MN/m^2) had scarcely changed and was exceeded by those of the sand and gravel concrete and of an expanded slag concrete. For the 31 MN/m^2 series, the values were relatively low, 4·0 and 3·9 MN/m^2 after 28 and 90 days, and were sixth and fifth respectively in the overall descending order for the materials that were compared.

Empirical relationships between modulus of rupture and compressive strength have been developed for concretes of various kinds. It is recognized that the modulus of rupture generally increases with increasing compressive strength but that it does so at a decreasing rate;[100,101] furthermore, a number of investigations have suggested that it is directly related to the square root of the compressive strength by empirical expressions of the form

$$f_r = K(f_c)^{1/2} \tag{1}$$

where f_r represents the modulus of rupture, f_c the compressive strength and K a coefficient.

Swamy, Ibrahim and Anand[92] found that a relationship of this kind could be applied to concretes made from the British manufactured expanded slate aggregate, both with and without partial sand replacement. With units expressed in N/mm^2, they found values of the coefficient to be 0·75 and 0·56, for wet and dry specimens respectively, under the experimental conditions which they employed. (They found a similar relationship between tensile splitting strength and compressive strength and, in this case, obtained values for the coefficient of 0·56 and 0·44 for wet and dry specimens.)

Bobrowski and Bardhan-Roy[95] pointed out that the value of the coefficient in such relationships decreases above a compressive strength of 34·5 MN/m^2 and gave values (adjusted here for these units) of 0·66 for use below this level and 0·42 above it; however, in the light of field tests, they suggested that a more realistic value in the latter case might be 0·54.

It has also been shown[102] that the modulus of rupture values for some lightweight aggregate concretes, with compressive strengths between 20 and 60 MN/m^2, generally fall close to the line represented by the expression

$$f_r = 0{\cdot}46(f_c)^{2/3} \tag{2}$$

with symbols as in eqn (1).

Various published modulus of rupture values and ranges of values, for expanded slate concretes at ages of at least 28 days, are plotted as a function of density in Fig. 3.17; and calculated values, obtained from the corresponding published compressive strength data using the expressions above, are plotted for comparison. A value of 0·67 was assumed for the coefficient used in eqn (1);[101] and this expression seems to provide a slightly better overall estimate of the measured values although, of course, the

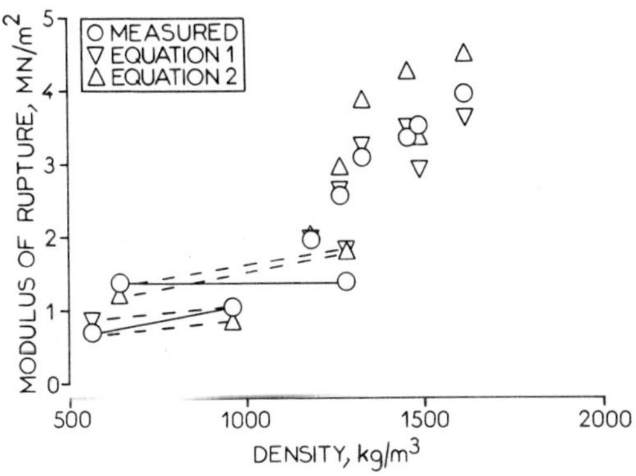

Fig. 3.17. The relationship between modulus of rupture (measured and estimated) and density of expanded slate concretes at ages of at least 28 days. (Data from refs. 38, 86 and 87.)

amount of data is very limited and has been obtained by different workers using different experimental conditions.

3.5.5. Elastic properties

Kluge, Sparks and Tuma[86] found that, in common with the other relatively hard aggregates such as the expanded clay, shale and slags which they examined, the elastic modulus of the expanded slate concretes tended to be comparatively high with values that were strongly dependent on the cement content. As the cement content was varied from about 170 to nearly 500 kg/m³ (with corresponding room dry densities rising from 1190 to 1460 kg/m³), the static modulus of elasticity varied from 7·6 to 12·0 GN/m² and the dynamic values from 8·6 to 13·1 GN/m². (By contrast they found that the softer aggregates gave concretes with very low elastic modulus values which were only slightly affected by the cement content, if at all.)

The two expanded slate mixes described by Shideler,[87] with densities of 1490 and 1620 kg/m³, gave 28 day static modulus of elasticity values of 12·8 and 15·1 GN/m² respectively which, after 365 days' moist curing, became

15·6 and 17·7 GN/m^2; the same mixes, moist cured for 7 days followed by storage at 50% relative humidity, gave corresponding values of 11·3 and 14·1 GN/m^2 after 28 days and 10·1 and 14·2 GN/m^2 after 365 days (see Fig. 3.18).

Swamy and Bandyopadhyay[91] published the results of an investigation into the elastic properties of concrete made with the British expanded slate

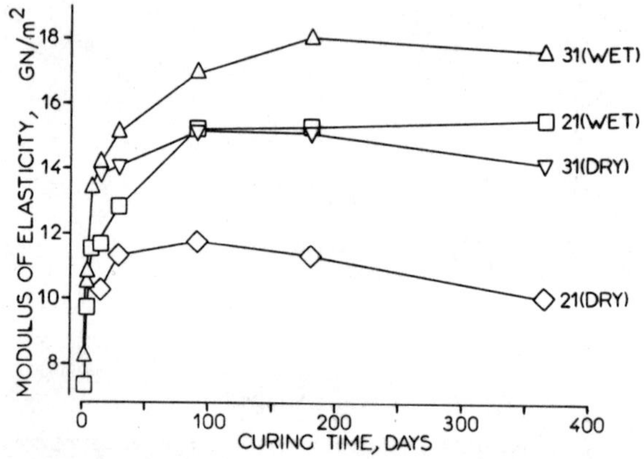

Fig. 3.18. The development of modulus of elasticity for expanded slate concretes with nominal compressive strengths of 21 and 31 MN/m^2, under wet (continuous moist curing) and dry (50% relative humidity after 7 days' moist curing) conditions (after Shideler[87]).

aggregate. They prepared five mixes, using ordinary Portland cement with coarse aggregate from 19 to 5 mm (bulk density about 720 kg/m^3) and fines from 4·8 mm downwards (bulk density about 1000 kg/m^3). The cement content was varied between 230 and 410 kg/m^3 and the 28 day compressive strengths ranged from 21 to 60 MN/m^2. After demoulding 24 hours from casting, they used four different curing conditions: 16 ± 0·5 °C and 50 ± 2% relative humidity; 27 ± 3 °C and 95 ± 5% relative humidity; uncontrolled laboratory conditions; and 7 days at 27 ± 3 °C and 95 ± 5% relative humidity followed by uncontrolled laboratory conditions. They measured compressive strength, static and dynamic modulus of elasticity, Poisson's ratio and the damping characteristics (expressed as the logarithmic decrement) at various times up to 2 years.

They found that both moduli of elasticity increased with compressive strength and their results suggested that the curing conditions used did not have a very significant effect on the values obtained, although some differences were noted. Increase in elastic modulus with age was generally more pronounced with moist curing and the air dried specimens tended to show a slight loss after a year (cf. Fig. 3.18) which was associated with differential moisture movement and shrinkage cracking resulting from air drying. Over the strength range investigated, 28 day modulus values varied from about 17 to 22 GN/m² and 19 to 26 GN/m² for static and dynamic determinations respectively. Using 28 day values, they related elastic modulus (E) to density (w) and compressive strength (f_c) by means of the expression

$$E = Kw^2(f_c)^{1/2} \qquad (3)$$

and found values for the coefficient, K, which appeared to correspond to values obtained by other workers using lightweight concretes. Using their own and other data, including some from different lightweight aggregates, they suggested that 28 day values of static modulus (E_s in kN/mm²) and dynamic modulus (E_d in kN/mm²) are related by the expression

$$E_s = 1{\cdot}04E_d \quad 4{\cdot}11 \qquad (4)$$

and that this is correct, within ± 4 kN/mm², for structural concretes made from British lightweight aggregates. They also provided comparative data showing that, over the range of compressive strengths from 21 to 60 MN/m², static modulus values for their expanded slate concretes were greater than those based on a sintered PFA and an expanded clay aggregate and, at low strengths, on a foamed slag although this latter material gave higher values at strengths of about 40 MN/m² and above.

They found that Poisson's ratio increased with age and that a very large proportion of its value at 1 year had been developed after 28 days; furthermore, the curing conditions generally appeared to have no significant effect. Poisson's ratio increased with compressive strength and, over the strength range used, 28 day values increased from about 0·19 to 0·22; and the concomitant increase in modulus led them to develop an empirical expression relating Poisson's ratio to static modulus and compressive strength.

The damping capacity of the expanded slate concretes decreased with age and was affected by the curing conditions; for moist cured specimens the rate of decrease in damping was relatively slow after 28 days but for air dried specimens it continued after a year. There was a general reduction in logarithmic decrement with increasing compressive strength at 28 days, values ranging from above 0·04 to just below 0 03 and, taking the results as a whole, values varied between 0·12 and 0·02. Swamy and Bandyopadhyay concluded that the damping capacity of their expanded slate concretes was comparable to that of gravel concrete.

Modulus of elasticity is an important property and there are a number of empirical expressions which are intended to relate its value, for lightweight concretes, to compressive strength and which take the density of the material into account.

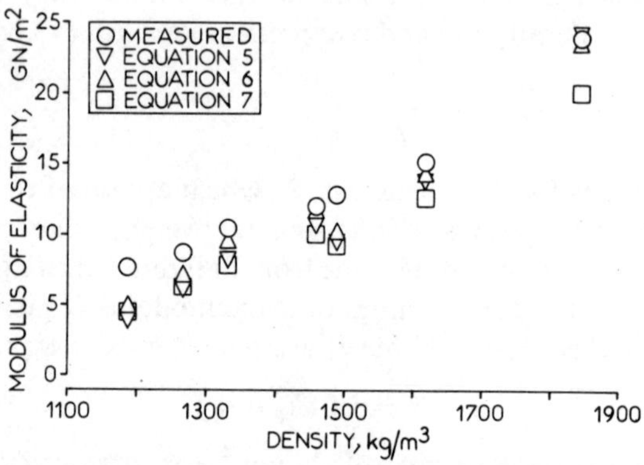

Fig. 3.19. The relationship between modulus of elasticity (measured and estimated) and density of some expanded slate concretes. (Data from refs. 86, 87 and 95.)

Swamy and Bandyopadhyay[91] and Bobrowski and Bardhan-Roy[95] have used the relationship given by eqn (3). For lightweight concrete based on the British manufactured expanded slate aggregate, this takes the form

$$E = (0{\cdot}97 \times 10^{-3})w^2(f_c)^{1/2} \tag{5}$$

when elastic modulus (E) and compressive strength (f_c) are expressed in MN/m^2 and density (w) in kg/m^3.

Pauw's formula[103] also relates the elastic modulus to the square root of the compressive strength (at a given age) and the density (air dry at 28 days) as follows:

$$E = 0{\cdot}04w^{3/2}(f_c)^{1/2} \tag{6}$$

Furthermore, Short and Kinniburgh[104] quote the relationship

$$E = (1{\cdot}58 \times 10^{-3})w^2(f_c)^{1/3} \tag{7}$$

Figure 3.19 shows published elastic modulus values, plotted against density, compared with corresponding estimates using each of the expressions above; the precise experimental conditions were not quoted for all the data but, nevertheless, Pauw's formula appears to give the best overall estimate of the actual values although, as with the modulus of rupture plot, the amount of data is very limited.

3.5.6. Shrinkage and creep

In comparing different aggregates in the form of precast concrete slabs, Coleman[38] found ranges of values for drying shrinkage, and expansion on wetting, that were generally markedly lower for expanded slate than for pumice-, breeze- and clinker-based concretes. Furthermore, the shrinkage values for the expanded slate concretes were considerably lower (generally less than half) than the maximum that was permitted by the British Standard Specification for precast concrete partition slabs.

Kluge, Sparks and Tuma[86] found that the expanded shale and slate concretes performed very well in drying shrinkage tests and showed the lowest average shrinkage values (823×10^{-6} and 838×10^{-6} m/m respectively after 100 days drying at 21 °C and 55 % relative humidity) of all the lightweight aggregates they investigated; values obtained from different mixes using the expanded slate aggregate ranged from 775×10^{-6} to 942×10^{-6} m/m, compared with between 400×10^{-6} and 625×10^{-6} m/m for the sand and gravel.

Of the seven lightweight aggregates that Shideler[87] used to prepare his 21 and 31 MN/m^2 nominal strength mixes, the expanded slate and one of the expanded shales gave the best results in comparative creep tests. In these tests, a load of 4·1 MN/m^2 was applied after 7 days' moist curing and the specimens were maintained at 23 °C and 50 % relative humidity; at the time of loading, the stress/strength ratios for the 21 MN/m^2 series varied between 0·26 and 0·37 (0·36 for the expanded slate concrete) and for the 31 MN/m^2 series between 0·15 and 0·20 (about 0·18 for the expanded slate concrete). At ages up to a year, the expanded slate concretes showed less creep than the corresponding sand and gravel mixes; at 1 year, the creep strain of the expanded slate concrete in the 21 MN/m^2 series was still slightly less than, but tending towards, that of the corresponding sand and gravel concrete but, in the 31 MN/m^2 series, there seemed to be no significant difference between the two aggregates. The specific creep values for the expanded slate materials were 170×10^{-6} and 120×10^{-6} m/m per MN/m^2 for the 21 and 31 MN/m^2 mixes respectively, at 1 year; the corresponding drying shrinkage over the same period was about 820×10^{-6} m/m for the 21 MN/m^2 material, and about 680×10^{-6} m/m for the sand and gravel concrete in the same series.

The 21 MN/m^2 series of materials were also subjected to creep tests after

7 days of moist curing followed by a further 21 days at 50% relative humidity; this resulted in similar but somewhat lower creep strains than those obtained by loading after 7 days' moist curing.

Swamy and Ibrahim[88] investigated the shrinkage and creep properties of concretes made with the British manufactured expanded slate aggregate and an ultra fine Portland cement (with a specific surface greater than 700 m^2/kg) that developed high early strength. They used two mixes, both with an aggregate/cement ratio of 2·0 by weight and a water/cement ratio of 0·55. The coarse aggregate (9–19 mm) and the fines had typical bulk densities of 620–720 and 980–1060 kg/m^3 respectively—and the fines, amounting to 44% of the total aggregate weight, consisted of all expanded slate in one mix and in the other, two-thirds of its weight was replaced by sand. After 28 days' moist curing, the compressive strength of these materials was 46 and 50 MN/m^2 respectively, with and without sand replacement. Various curing regimes were used, generally involving moist curing (27 ± 3 °C and 95 ± 5% relative humidity) and controlled temperature and humidity conditions ($15{\cdot}7 \pm 0{\cdot}5$ °C and 50 ± 2% relative humidity) in different combinations.

The drying shrinkage of these materials at 600–700 days ranged between 635×10^{-6} and 830×10^{-6} m/m depending on the conditions employed; for instance, after 28 days' moist curing, drying at $15{\cdot}7 \pm 0{\cdot}5$ °C and 50 ± 2% relative humidity gave a shrinkage of 720×10^{-6} m/m at 686 days for the all-slate concrete containing no sand. Swamy and Ibrahim commented that their results compared well with Shideler's and with those from other lightweight concretes; the shrinkage was about 10–15% less than that obtained with sintered PFA and expanded clay concrete in other investigations but 12–20% more than with some American lightweight aggregate concretes.

Their creep tests employed stress/strength ratios generally between about one-third and one-half (and creep was believed to vary linearly with stress/strength ratio up to some value between these levels). Specific creep values ranged between 75×10^{-6} and 105×10^{-6} m/m per MN/m^2 at ages of 560–660 days for materials loaded at 28 days; for instance, one test, corresponding to the shrinkage example quoted in the previous paragraph and using a stress/strength ratio of 0·35 at 28 days, gave a specific creep value of 74×10^{-6} m/m per MN/m^2 at 659 days. The creep of these concretes was found to be comparable to that obtained from a sintered PFA concrete made with a rapid hardening cement, but 20–37% less than that from expanded clay—and it was found to be up to 20% more than that from a gravel concrete made with the same fine cement.

Swamy and Ibrahim found that the initial shrinkage and creep rates were very high with the ultra fine Portland cement; in both cases, a very large proportion of the deformation had occurred within a year. Furthermore, neither shrinkage nor creep generally appeared to be significantly affected by partial replacement of the expanded slate fines with sand, although some differences were noted. From their investigation, they developed shrinkage and creep data for design purposes, using the particular expanded slate they employed together with the ultra fine Portland cement, and this information can be found in their paper. (In a later publication, Swamy[89] compared the shrinkage results with those obtained, under similar conditions after 700 days, from concrete based on a granite aggregate; he concluded that, using finely ground cement, the type of aggregate appeared to have little effect upon the value of the final shrinkage under the experimental conditions that he employed.)

Bobrowski and Bardhan-Roy[95] used camber readings from expanded slate concrete roof beams to obtain estimated specific creep values, at 6 months, ranging from between 46×10^{-6} and 78×10^{-6} m/m per MN/m^2, with an average of nearly 73×10^{-6} m/m per MN/m^2. The concrete was based on an aggregate/cement ratio of 2·1 by weight, with 60% coarse expanded slate and 40% sand by weight, and the cement content was 534 kg/m^3; the total water/cement ratio was 0·39 and a plasticizer was used. Similar measurements, with sintered PFA concrete roof beams, indicated considerably greater specific creep values, with an average of about 91 m/m per MN/m^2; in this case the cement content was slightly higher (578 kg/m^3), with a slightly lower aggregate/cement ratio (1·95 by weight), but less sand was used (31% of the total aggregate weight) and the total water/cement ratio was higher (0·43).

Short[99] quoted a range of shrinkage values of 300×10^{-6} to 900×10^{-6} m/m for expanded slate concretes, based on the British manufactured aggregate, with corresponding densities ranging from 1840 to 800 kg/m^3 and compressive strengths from 62 to 1·4 MN/m^2; however, the test conditions were not specified. And Neville[98] quoted 400×10^{-6} and 450×10^{-6} m/m for densities of 1700 and 1750 kg/m^3 and compressive strengths of 28 and 35 MN/m^2 respectively, also without specifying the test conditions.

Shideler[87] showed that, for both an expanded clay concrete and a sand and gravel concrete, there is a marked relationship between the ultimate creep value and the compressive strength at the time of loading; the ultimate creep fell rapidly as the compressive strength increased up to about 45 MN/m^2, but above this value it fell more slowly.

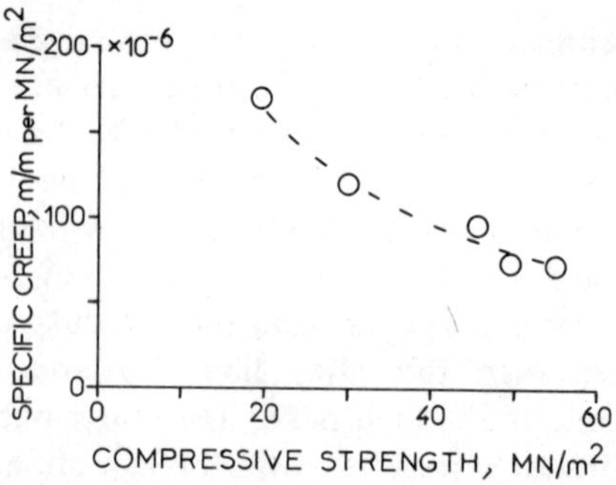

Fig. 3.20. The relationship between specific creep and compressive strength for some expanded slate concretes. (Data from refs. 87, 88 and 95.)

Figures 3.20 and 3.21 show both specific creep and shrinkage values for expanded slate concretes, taken from sources quoted above, plotted against compressive strength; and both properties show a marked reduction with increasing strength although the amount of data is very limited. Obviously some caution should be exercised in interpreting data collated from different sources since test methods vary considerably; and this can be particularly true for creep and shrinkage testing where a wide range of ambient conditions are used for both curing and testing, as well as all the other variables involved. However, apart from Bobrowski and Bardhan-Roy's field test result, the laboratory creep and shrinkage data used in the figures was obtained from specimens which had been initially moist or fog cured, for either 7 or 28 days, followed by testing at normal temperatures (between about 16 and 23 °C) and relative humidities of about 50–55 %; the laboratory tests generally extended over a period of a year or more

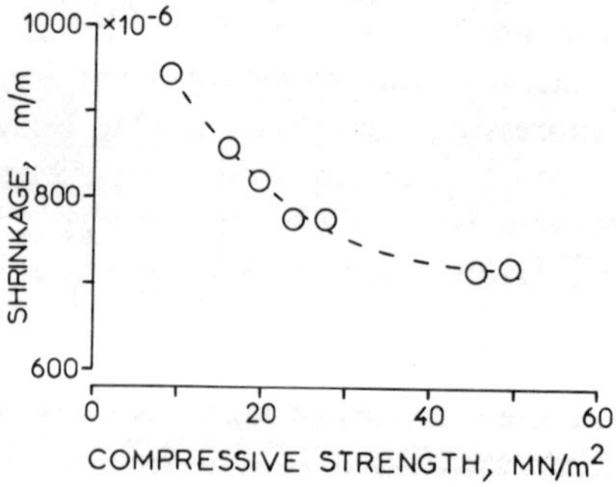

Fig. 3.21. The relationship between shrinkage and compressive strength for some expanded slate concretes. (Data from refs. 86, 87 and 88.)

(although the shrinkage values obtained by Kluge, Sparks and Tuma were taken after only 100 days, but they used high early strength cement for their mixes). The laboratory creep tests were performed at stress/strength ratios of less than 0·4, and 28 day values were used for the compressive strength.

3.5.7. Water absorption

The porous nature of lightweight aggregates means that their water absorption values tend to be high in comparison with conventional dense aggregates; and this affects the properties of the hardened concretes from which they are made. But, as noted earlier, it also affects the approach to mix design; furthermore, it can give rise to problems, with the fresh concrete, related to its workability and also to the ease with which it can be pumped.

The pressure involved in pumping concrete tends to force water into the porous lightweight aggregate particles and the mix then becomes stiffer and more difficult to pump; this problem can be reduced by various expedients such as presaturating the aggregate, modifying the mix design and mixing technique, and using fine sand and other additives.[105,106] But even after the pumping operation is finished, when the pressure has returned to normal, water migrating out of the aggregate particles can adversely affect the bond around them.[106] Obviously for pumping operations, as well as for other reasons, relatively low water absorption is a desirable characteristic in a lightweight aggregate. The expanded slate aggregate that was produced in the United Kingdom had relatively low water absorption values compared with some other British manufactured lightweight aggregates.[107-109] And this factor contributed towards its choice for one application in the United Kingdom, where lightweight concrete was needed because of foundation loading restrictions and where pumping was desirable because of access problems.[110]

The water absorption of lightweight slate aggregate depends upon the extent to which it has been expanded and, generally, absorption values rise with decreasing density; but, superimposed on this relationship, there are effects arising from the mineralogical and chemical constitution of the raw material and from the processing conditions employed. The measured absorption value depends very much on the technique used to determine it. For instance, Landgren[111] investigated the water absorption characteristics of expanded slate from the same stock employed by Shideler and found that, after 30 min immersion, only a relatively small proportion of the potential value was reached for both fine (No. 8–No. 4 sieve sizes) and coarse (6·4–9·5 mm) material. He found absorptions of 5·2 and 4·9 % of the aggregate dry weight, for fine and coarse fractions respectively, after 30 min immersion whereas after 1 day the corresponding values had risen to 9·9 and 8·7 %; and after vacuum saturation followed by 5–7 weeks' immersion the values were 20·2 and 28·3 % respectively.

In practice, water absorption is often determined by initially weighing the dry aggregate (in the oven-dried condition, for example) and then weighing it again after 24 hours' immersion in water; the absorption value, which corresponds to the difference between the weights, can be expressed as a percentage of the dry weight although sometimes it is quoted on a volume basis either in relation to the material itself or to its bulk volume including the gaps between the aggregate particles. But even such an apparently straightforward procedure can present difficulties with lightweight aggregates, particularly in determining the weight after immersion. Ideally, the sample is weighed in the saturated surface-dry condition, i.e. with the surface water removed (using a towel, for example), but care must be taken to remove all the surface water without removing a significant amount from the porous structure within the aggregate particles. The problems of determining the water absorption of lightweight aggregates have been considered in some detail in another paper by Landgren,[112] who studied the properties of fifteen coarse lightweight aggregates including an expanded slate, and in a discussion by Lydon.[113]

Different methods of determining and expressing water absorption values and density values of the aggregate sometimes make it difficult to directly compare data quoted from different sources in the literature. But Fig. 3.22 shows 24 hour absorption values, expressed by weight, plotted against density values obtained from one source, namely the report by Hamlin and Templin;[76] although wide ranges of values are obtained at similar densities, it is clear that the more expanded materials tend to absorb greater quantities of water.

It should be noted that lightweight aggregates 'adsorb' water vapour from the atmosphere even though they may not be in actual contact with the liquid; such adsorbed water molecules are strongly attracted to the surfaces of the material, both internal and external, by physicochemical forces—and the quantity of adsorbed water increases with the ambient relative humidity. Landgren[111] investigated the desorption from saturation, in steps, followed by the adsorption of water vapour by fifteen different lightweight aggregates, including the expanded slate from the same stock used by Shideler. Figure 3.23 shows the desorption–adsorption characteristics of the expanded slate fine and coarse fractions expressed as the quantity of sorbed water at equilibrium at various relative humidities; clearly, the

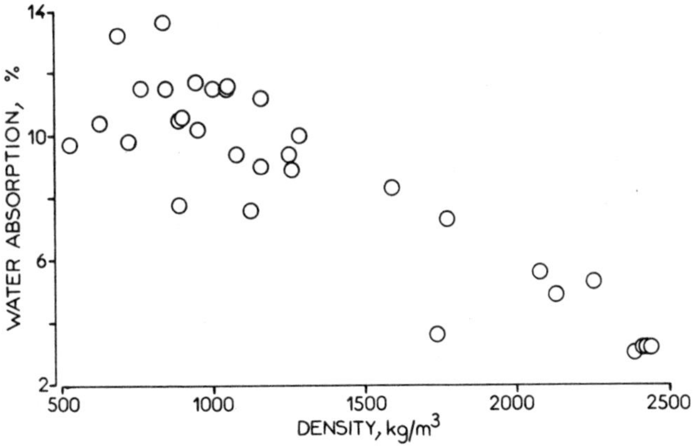

Fig. 3.22. The relationship between 24 hour water absorption, by weight, and density of some expanded slate aggregates (after Hamlin and Templin[76]).

quantities of water involved are considerably less than those absorbed by immersion in liquid water. Similar measurements were made with fourteen other lightweight aggregates including expanded clays, shales and slags as well as pumices, volcanic cinders and soft coal cinders. Landgren also calculated the internal specific surface area of the lightweight aggregates from these results using two different methods and considered that the BET equation (from the theory of multilayer adsorption by Brunauer, Emmett and Teller) gave the most reliable values. An expanded shale and the expanded slate aggregates gave the lowest internal specific surface areas

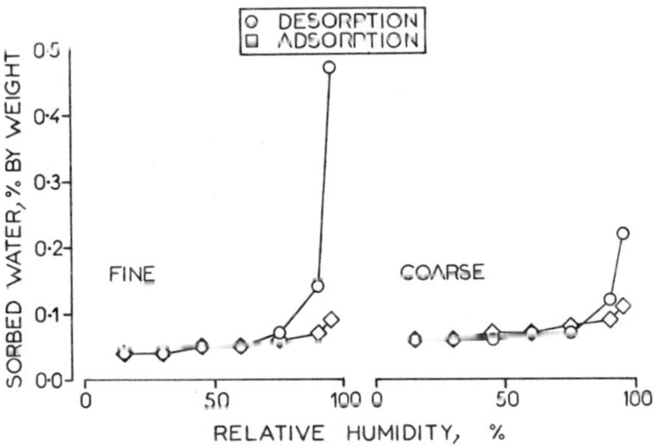

Fig. 3.23. Water vapour desorption–adsorption characteristics for fine and coarse fractions of an expanded slate aggregate (after Landgren[111]).

which, in the latter case, were 1000 and 1400 m^2/kg for the fine and coarse material respectively.

Landgren compared the values for lightweight aggregates with those obtained from some dense, natural aggregates used for concrete; apart from orthoquartzite, with a BET specific surface area of 100 m^2/kg, the slate aggregate gave lower values than all these, which ranged from 3000 to 7000 m^2/kg for materials including granite, gabbro, dolomite and

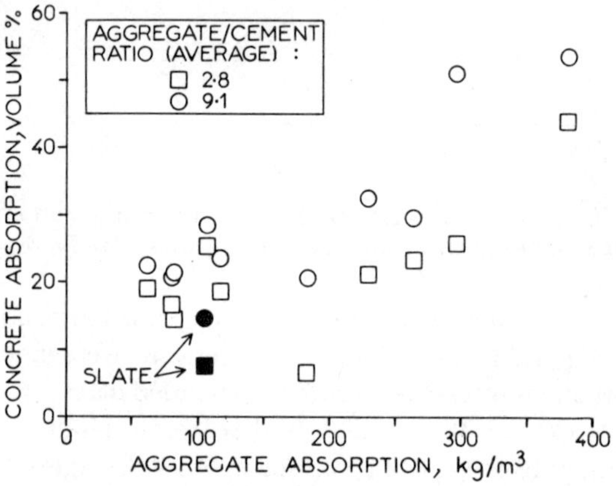

Fig. 3.24. The relationship between concrete and aggregate water absorption values for different lightweight aggregates (after Kluge, Sparks and Tuma[86]).

greywacke. And with its considerably reduced bulk specific gravity, the specific surface area of expanded slate on a volume basis was even more markedly lower.

Landgren discussed the results of his investigation and their application in the context of the performance of lightweight aggregates in concrete; in particular, he referred to the drying of concrete masonry products and to the effect of moisture content on the fire resistance of concrete and on its freeze–thaw durability.

The water absorption of a lightweight aggregate is reflected in the absorption values of concretes in which it is used but, again, the proper comparison of data from different sources can be difficult. Figure 3.24 shows the relationship between concrete absorption and aggregate absorption for the different lightweight aggregates investigated by Kluge, Sparks and Tuma;[86] the concrete absorption is expressed in terms of volume, and the aggregate absorption, which in effect can be viewed in the same way, has been derived from their data to take into account the different absorption values obtained from the coarse and fine fractions which were used in various proportions. Data for the highest and lowest aggregate/cement ratios are shown, i.e. those series of mixes with average ratios of 9·1 and 2·8 respectively. Although the scatter reflects the effect of other factors not taken into account, it is clear that the more absorbent aggregates tend to give more absorbent concretes and, as might be expected, the lower aggregate/cement ratio gives lower concrete absorption values. The expanded slate concretes have particularly low absorption values that are below the values for the corresponding sand and gravel concretes of 15·0% and 9·3% with aggregate/cement ratios of 10·0 and 3·2 respectively.

3.5.8. Durability

Bandyopadhyay and Swamy[90] published some interim results from a long-term study of the durability of steel embedded in structural lightweight concrete based on the British manufactured expanded slate aggregate. Concretes had been prepared using the expanded slate aggregate alone and using it with the replacement of two-thirds of the lightweight fines by natural sand—and mixes based on a gravel aggregate were also included. For each type of aggregate, four mixes were prepared to produce concretes of four nominal compressive strength levels, namely 21, 31, 41 and 52 MN/m^2 at 28 days (with cement contents of 230, 260, 360 and 400 kg/m^3 respectively for the expanded slate concretes)—and the specimens contained steel bars with the cover ranging from 13 to 25 mm. After 7 days' moist curing and a further 21 days under uncontrolled laboratory conditions, the specimens were exposed to the weather at a site noted for a high degree of atmospheric pollution and with a high content of sulphur in the air.

After 1 and 2 years' exposure, both series of expanded slate concretes suffered surface crazing which increased markedly for the stronger materials, and was particularly obvious for the strongest (i.e. nominally 52 MN/m^2 at 28 days); Bandyopadhyay and Swamy associated this crazing with the relatively high proportion of fines that was used.

They measured the depth of carbonation, using phenolphthalein indicator on broken surfaces, after 1 and 2 years; and 24 hour alcohol absorption tests indicating relative permeability, which is highly correlated with carbonation depth, were carried out after 2 years. Alcohol permeability and the depth of the carbonation front were markedly reduced by increasing cement content and by the partial replacement of the lightweight fines by sand. Figure 3.25 shows the alcohol permeability results. No depth of carbonation was detected for the two strongest all-slate concretes (with cement contents of 360 and 400 kg/m^3) or for the three strongest with partial sand replacement (with cement contents of 260, 360 and 400 kg/m^3); the maximum recorded was 8–12 mm for the weakest all-slate material after 2 years' exposure. It was also noted that large pieces of expanded slate aggregate could provide a 'short-cut' for progress of the carbonation front but that no damage due to corrosion of the steel bars was detected on inspection after 2 years' exposure.

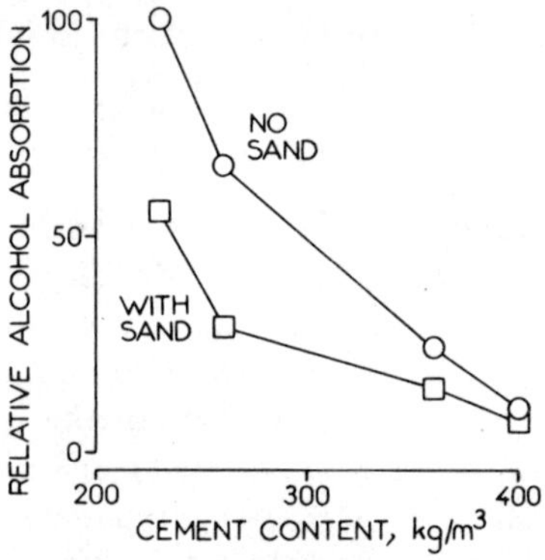

Fig. 3.25. The effect of cement content and partial replacement of lightweight fines by sand on the alcohol permeability of expanded slate concrete after 2 years' exposure (after Bandyopadhyay and Swamy[90]).

A different aspect of durability was investigated by Kluge, Sparks and Tuma;[86] they carried out freezing and thawing tests with a cycle ranging from freezing in air at −46 °C for 4 hours to thawing in water at about 16 °C for about 15 hours. The expanded slate concretes withstood this treatment extremely well and, together with the sintered fly ash concretes, were the best of all the materials investigated in this respect; and it appeared that this behaviour was related to the low water absorption of the concrete.

3.5.9. Thermal conductivity

The thermal conductivity of expanded slate concrete is clearly a very important property at a time when the thermal insulation performance of constructional materials is assuming increasing significance. However, there is relatively little information in the literature. Some published values and ranges of values are plotted against density in Fig. 3.26.

The performance of expanded slate concrete, relative to other lightweight aggregate concretes, seems to depend upon density. For instance, Spratt[114] has provided data comparing the thermal conductivities of various lightweight aggregate concretes with densities of 1500–1600 and 950–1050 kg/m^3. For the heavier concretes, the value obtained from the British manufactured expanded slate aggregate (0·68 W/m °C) is markedly higher than the corresponding values obtained from other lightweight aggregates, including clinker, expanded clays, sintered PFA and foamed slag, which range from between 0·39 and 0·58 W/m °C. On the other hand,

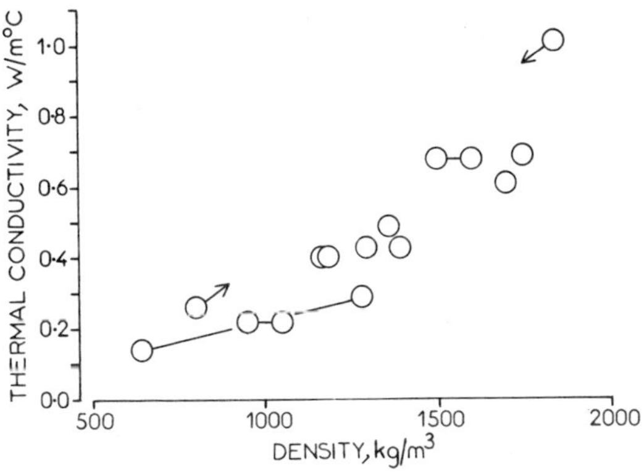

Fig. 3.26. The relationship between thermal conductivity and density of some expanded slate concretes. (Data from refs. 25, 26, 86, 98 and 99.)

for the 950–1050 kg/m^3 concretes, the value from the expanded slate aggregate (0·22 W/m °C) is towards the bottom of the overall thermal conductivity range which extends from 0·16 to 0·35 W/m °C. Furthermore, Kluge, Sparks and Tuma[86] found that expanded slate concretes, compared with the others they investigated in the same density range (1170–1390 kg/m^3 in the oven dry condition), tended to have relatively high thermal conductivity values ranging from 0·40 to 0·49 W/m °C although these are very similar to those from an expanded clay. Nevertheless, the values were very much less than those from the corresponding sand and gravel mixes which ranged from nearly twice to well over three times as high. (The thermal conductivity of ordinary dense concrete has been given[114] as 1·15–1·44 W/m °C for densities of 2240–2400 kg/m^3.)

It should be noted that the high thermal conductivity of water, compared with that of stationary air, means that the insulation properties of lightweight aggregate concretes are very much dependent on their moisture content. Both Spratt[114] and Short and Kinniburgh[115] provide data which show that the thermal conductivity of dry concrete is considerably increased even by relatively small moisture contents; for instance, moisture contents of 1 %, 5 % and 25 % lead to an increase in thermal conductivity of 30 %, 75 % and 175 % respectively. So the thermal conductivity of lightweight aggregate concrete depends on its moisture content under given ambient conditions; and the relatively low adsorption and absorption characteristics that some expanded slates have been shown to possess might well prove to be advantageous in this respect.

It has been pointed out[116] that the lower thermal conductivity and the lower mass (hence reduced heat capacity) of lightweight concrete, compared with dense aggregate concrete, tend to result in relatively high internal temperature rises after casting due to the heat liberated by the hydration of the cement. Furthermore, relatively high cement contents are often used in lightweight aggregate concretes and corresponding amounts of heat are evolved. Owens[117] measured the temperature rise, due to hydration, within insulated 300 mm cubes made from various concretes using different aggregates with a constant cement content of 440 kg/m^3 (and water contents varying between 185 and 300 litre/m^3). An expanded slate aggregate gave a more rapid initial temperature rise than other lightweight aggregates investigated (including a sintered PFA, two expanded clays and a foamed slag) as well as a dense aggregate; furthermore, it gave the second highest maximum temperature. Such effects can be advantageous in reducing the risk of frost damage in cold weather[116] but the influence of temperature on strength development is complex[118,119] and can be adverse.

3.5.10. Fire resistance

It has been pointed out that many lightweight aggregate concretes have better fire resistance than normal dense concrete.[120,121] The higher void content and relatively low coefficient of thermal expansion, as well as the fact that many lightweight aggregates are manufactured at high temperatures, contribute to this resistance. Spalling tends to be reduced and loss of strength can be markedly lower; furthermore, the lower thermal conductivity tends to afford better protection of steel reinforcement. Spratt[121] noted that, for some lightweight concretes with high moisture contents, the internal temperature rise within the material may be significantly delayed due to the absorption of heat energy involved in evaporating the water. It appears that spalling is also affected by the moisture content.[122]

Spratt[121] also provided data indicating how the thickness of concrete required for a given notional period of fire resistance may be reduced by using expanded slate concrete (based on the British manufactured aggregate) as well as other lightweight aggregate concretes, compared with dense aggregate concrete, particularly for floor slabs.

3.5.11. Miscellaneous properties

Various other properties of expanded slate concretes have been mentioned in the literature. For instance, Coleman[38] found that expanded slate concrete performed well in sawing and nailing tests; and the force needed to extract nails was twice that for a typical pumice concrete. Plastering tests, using cement, lime and gypsum plasters, gave good results and the material provided an excellent key; adhesion tests showed that the plaster failed before the bond.

Shideler[87] reported the results of pull-out tests (ASTM C 234-49 T) that were used to assess bonding between reinforcing steel and the concretes which he investigated; the expanded slate materials gave bond strength values of 4·2–5·9 and 5·3–6·3 MN/m^2, for materials with nominal compressive strengths of 21 and 31 MN/m^2 respectively, depending on the conditions of the test. He also found that none of the aggregates which he investigated, including the expanded slate, showed evidence of alkali–aggregate reaction after storage of concrete specimens for 9 months over water at 38 °C; the cement he used had an alkali content of 0·63 (expressed as Na_2O).

Finally, there have been some recent publications concerned with reinforced and prestressed concrete using expanded slate aggregate,[93,94,109] and there has also been an investigation into sound insulation properties.[123]

REFERENCES

1. W. Gutt, P. J. Nixon, M. A. Smith, W. H. Harrison and A. D. Russell, *A Survey of the Locations, Disposal and Prospective Uses of the Major Industrial By-products and Waste Materials*, Building Research Establishment Current Paper CP 19/74, Department of the Environment (Feb. 1974), Chapter 3.
2. F. M. Lawson and P. J. Nixon, *A Survey of the Locations, Disposal and Prospective Uses of the Major Industrial By-products and Waste Materials in Scotland*, Building Research Establishment Current Paper CP 50/78, Department of the Environment (Jun. 1978).
3. K. Akerman, B. Żmudziński, Z. Orman and Z. Musienko, *Arch. Hutnictwa*, **1** (1956), 319–39. (*Chem. Abstr.*, **51** (1957), 12383b.)
4. K. L. Watson, R. D. Grant and R. M. Jones, 'A potential use for slate waste?', *Precast Concrete*, **7**(9) (Sep. 1976), 449–52.

5. A. M. Neville, *Properties of Concrete*, 2nd (metric) Edition, Pitman Publishing, London (1973), 237–9.
6. K. L. Watson, R. D. Grant and R. M. Jones (Department of Civil Engineering, Portsmouth Polytechnic), unpublished work (1975–6).
7. R. H. Boyle (Director, Penrhyn Quarries Ltd), private communication.
8. I. Schmid, *Silikat Tech.*, **12** (1961), 22–4. (*Chem. Abstr.*, **55** (1961), 18055b.)
9. *Report of Aggregates and Waste Materials Working Group*, Building Research Establishment Current Paper CP 31/73, Department of the Environment (Nov. 1973), 5.
10. W. H. Gutt and M. A. Smith, 'Aspects of waste materials and their potential for use in concrete', *Resource Recovery and Conservation*, **1** (1976), 345–67.
11. J. Setchell, 'The Old Delabole Slate Quarries, Cornwall', *The Quarry Managers' Journal*, **16**(3) (Jun. 1933), 93–7.
12. F. D. Hoyt, *What's New in the Pennsylvania Slate Industry*, Mineral Industries Experiment Station Bulletin No. 66, The Pennsylvania State University (1956), 21–3.
13. H. O'Neill, 'The Dinorwic Slate Quarry at Llanberis', *The Quarry Managers' Journal*, **43**(12) (Dec. 1959), 469–73.
14. *The DORIC Range of Slate Facing Bricks*, Concrete Masonry Group, Wrexham (Mar. 1977).
15. K. H. McGhee and M. A. Ozol, *An Evaluation of a Slate Aggregate for Use in Concrete*, Virginia Highway Research Council (Nov. 1966).
16. K. H. McGhee (Senior Research Scientist, Commonwealth of Virginia Highway and Transportation Research Council), private communication.
17. N. H. Fox (Central Electricity Generating Board), private communication.
18. *Specification for Road and Bridge Works*, Ministry of Transport, Scottish Development Department and the Welsh Office, HMSO, London (1969), Table 13 (p. 49).
19. *North and Mid Cornwall Water Board*, *Crowdy Reservoir*, Lemon & Blizard (Consulting Engineers), Plymouth; also J. D. Evans (Resident Engineer), private communication.
20. Ref. 5, 522ff.
21. C. Forder, 'Lightweight concrete's place in the insulation spectrum', *Concrete*, **9**(1) (Jan. 1975), 28–30.
22. *Lightweight Concrete: History, Applications, Economics*, Expanded Shale Clay and Slate Institute, Washington, D.C. (1971).
23. B. H. Spratt, *An Introduction to Lightweight Concrete*, Cement and Concrete Association, London (1975).
24. Portland Cement Association, *Special Concretes, Mortars and Products*, John Wiley and Sons, Inc., New York (1975), Chapters 1 and 2.
25. A. Short and W. Kinniburgh, *Lightweight Concrete*, 3rd Edition, Applied Science Publishers Ltd, London (1978).
26. B. H. Spratt, *The Structural Use of Lightweight Aggregate Concrete*, Publication 45.023, Cement and Concrete Association, London (Dec. 1974), 64–5.
27. 'Lightweight aggregate concrete', *CEB/FIP Manual of Design and Technology*, Comité Euro-International du Béton/Fédération Internationale de la Précontrainte, The Construction Press Ltd, Lancaster (1977), 19–21.

28. ibid., 3.
29. Ref. 22, 44.
30. H. C. Robinson (Managing Director, Expanded Shale Clay and Slate Institute), private communication.
31. H. N. McCarl, 'Construction materials: aggregates—lightweight aggregates', *Industrial Minerals and Rocks* (*Nonmetallics other than Fuels*), 4th Edition (Editor-in-Chief: S. J. Lefond), American Institute of Mining, Metallurgical, and Petroleum Engineers, Inc., New York (1975), 85–96.
32. 'Lightweight aggregates—output of expanded clay, shale, and slate continues rise', *Rock Products*, **73**(12) (Dec. 1970), 59–60.
33. Ref. 22, 16.
34. Russ. 19,516 (1931). (*Chem. Abstr.*, **25** (1931), 3147.)
35. E. H. Coleman and P. J. Nixon, *A Survey of Possible Sources in Wales of Raw Materials for the Manufacture of a Lightweight Expanded Slate Aggregate*, Building Research Establishment Current Paper CP 78/74, Department of the Environment (Aug. 1974).
36. *The Utilization of Slate and Slate Waste*, Building Research Station Note No. A 84, Department of Scientific and Industrial Research (Jun. 1960).
37. E. H. Coleman, *Improvements in Cellular Products obtained from Laminated Rocks and the Like*, British Patent 401,685 (1933).
38. E. H. Coleman, 'A lightweight aggregate produced from slate waste', *Concrete and Constructional Engineering*, **31**(1) (Jan. 1936), 47–51.
39. C. R. Platzmann, *Betonstein-Ztg.*, **6** (1940), 55–6. (*Chem. Abstr.*, **36** (1942), 3925^3.)
40. *Properties and New Uses of Pennsylvania Slate*, Mineral Industries Experiment Station Bulletin No. 47, Pennsylvania State College (1947).
41. J. E. Conley, 'Waste slate as a raw-material source of lightweight aggregates', *Transactions of the American Institute of Mining, Metallurgical, and Petroleum Engineers*, **148** (1942), 161–6.
42. J. E. Conley, H. Wilson, T. A. Klinefelter *et al.*, *Production of Lightweight Concrete Aggregates from Clays, Shales, Slates, and other Materials*, Bureau of Mines Report of Investigations No. 4401, United States Department of the Interior (Nov. 1948).
43. D. Mocine, 'Lightweight aggregate from expanded slate', *Rock Products*, **52**(7) (Jul. 1949), 101–3, 116.
44. W. B. Lenhart, 'Automatically controlled kilns and fuel system for lightweight aggregate plant', *Rock Products*, **57**(5) (May 1954), 83–4, 86.
45. W. B. Lenhart, 'Producing lightweight aggregate from slate', *Rock Products*, **57**(3) (Mar. 1954), 92–6, 98.
46. W. E. Trauffer, 'Hudson's lightweight slate aggregate plant', *Pit and Quarry*, **56**(5) (Nov. 1963), 124–5, 128, 131, 136.
47. H. R. McCarl, 'Pennsylvania lightweight aggregate industry', *Minerals Processing*, **5**(12) (Dec. 1964), 18, 20, 25.
48. W. E. Trauffer, *Pit and Quarry*, **59**(8) (Feb. 1967), 105–8.
49. 'Lightweight concrete review. Part 2: production of low-density aggregates', *Concrete*, **2**(5) (May 1968), 190–6.
50. J. E. Conley and J. A. Ruppert, 'Recent developments in the manufacture of lightweight aggregates', *Mining Engineering* (Apr. 1950), 479–85.

51. US 2,456,207 (1948). (*Chem. Abstr.*, **43** (1949), 1548b.)
52. US 2,463,994 (1949). (*Chem. Abstr.*, **43** (1949), 3989d.)
53. US 2,654,136 (1953). (*Chem. Abstr.*, **48** (1954), 343b.)
54. US 2,654,137 (1953). (*Chem. Abstr.*, **48** (1954), 343c.)
55. *Improved Technique for Producing Expanded Aggregates for Use in Lightweight Concrete Products*, Building Research Station Note No. A 81, Department of Scientific and Industrial Research (May 1960).
56. Fr. 1,428,939 (1966). (*Chem. Abstr.*, **67** (1967), 76025r.)
57. Fr. 1,434,462 (1966). (*Chem. Abstr.*, **65** (1966), 18313f.)
58. Ger. Offen. 2,322,801 (1973). (*Chem. Abstr.*, **80** (1974), 73783u.)
59. Ger. Offen. 2,323,689 (1973). (*Chem. Abstr.*, **80** (1974), 86828n.)
60. Fr. 1,484,561 (1967). (*Chem. Abstr.*, **68** (1968), 32937f.)
61. P. C. Aitcin and C. Poulin, *Miner. Waste Util. Symp., 4th, Proc., Chicago, Ill.*, 7–8 May 1974, 265–8. (*Engng Index*, **74** (1975), 014372.)
62. N. W. Biege and S. M. Cohen, 'Cut fuel costs in the lightweight aggregate industry', *Ceramic Bulletin*, **54**(6) (1975), 569–70.
63. W. Davies, *The Utilization of Slate Waste*, Report prepared for The Development Board for Rural Wales (Mar. 1978).
64. F. D. Hoyt, 'Utilization of Pennsylvania slate for expanded aggregate', *Mining Engineering*, **10**(8) (Aug. 1958), 874–6.
65. R. Ironman, 'Thermoshock treatment produces expanded aggregates in 45 seconds', *Rock Products*, **74**(6) (Jun. 1971), 60–2.
66. G. Baudet, 'Les Matières premières expansibles', *Bull. Bur. Rech. Geol. Minières*, Section II, No. 5 (1971), 1–8. (*Chem. Abstr.*, **77** (1972), 51932c.)
67. F. Moeller, *Tonind.-Ztg. Keram. Rundsch.*, **92**(12) (1968), 498–501. (*Chem. Abstr.*, **70** (1969), 60396p.)
68. S. K. Das and W. B. Ledbetter, 'Variations of fundamental properties of synthetic aggregates with firing temperature', *The Pakistan Engineer*, **7**(3) (Mar. 1967), 195–206.
69. H. Aurich, 'Constructief licht beton, samengesteld met kunstmatig geëxpandeerde toeslagmaterialen', *Cement* (*Holland*), **19**(4) (1967), 120–7.
70. G. I. Konoshenko, V. K. Mukhina, G. A. Petrikhina and A. A, Akhundov, *Sb. Tr., Gos. Vses. Nauchno-Issled. Inst. Stroit. Materialov i Konstruktsii* (27) (1973), 88–98. (*Chem. Abstr.*, **82** (1975), 34476t.)
71. E. G. Ehlers, 'The mechanism of lightweight aggregate formation', *Ceramic Bulletin*, **37**(2) (1958), 95–9.
72. B. N. Vinogradov and L. E. Vrublevskii, *Sb. Tr., Gos. Vses. Nauchno-Issled. Inst. Stroit. Materialov i Konstruktsii* (5) (1965), 49–65. (*Chem. Abstr.*, **63** (1965), 17651f.)
73. W. H. Harrison, 'Synthetic aggregate sources and resources', *Concrete*, **8**(11) (Nov. 1974), 41–4.
74. M. Kalwa and H. Ropska, *Cement-Wapno-Gips*, **21**(11) (1966), 317–21. (*Chem. Abstr.*, **67** (1967), 5486f.)
75. Ger. Offen. 2,315,844 (1973). (*Chem. Abstr.*, **80** (1974), 99490w.)
76. H. P. Hamlin and G. Templin, *Evaluating Raw Materials for Rotary-kiln Production of Lightweight Aggregate*, Bureau of Mines Information Circular 8122, United States Department of the Interior (1962).
77. J. A. Hanson, 'American practice in proportioning lightweight-aggregate

concrete', *Proceedings of the First International Congress on Lightweight Concrete, London*, May 1968, The Concrete Society, London, Volume 1 (papers), 39–54.
78. Ref. 5, 534–41.
79. W. W. Hotaling, Jr, 'Using the recommended practice for selecting proportions for structural lightweight concrete (ACI 211.2-69)', *Proportioning Concrete Mixes*, Publication SP-46, American Concrete Institute, Detroit (1974), 209–19.
80. B. W. Shacklock, *Concrete Constituents and Mix Proportions*, Cement and Concrete Association, London (1974), 63–7.
81. A. Short, 'Lightweight aggregate concrete: Part 2', *Concrete*, **8**(8) (Aug. 1974), 41–2.
82. Ref. 25, Chapter 3.
83. Ref. 26, 18–22.
84. Ref. 27, 23–47.
85. P. L. Owens, 'Mix-design charts for lightweight-aggregate concrete', *Proceedings of the First International Congress on Lightweight Concrete, London*, May 1968, The Concrete Society, London (1970), Volume 2 (discussion), 24–5.
86. R. W. Kluge, M. M. Sparks and E. C. Tuma, 'Lightweight-aggregate concrete', *Journal of the American Concrete Institute*, **20**(9) (May 1949), 625–42.
87. J. J. Shideler, 'Lightweight-aggregate concrete for structural use', *Journal of the American Concrete Institute*, **29**(4) (Oct. 1957), 299–328.
88. R. N. Swamy and A. B. Ibrahim, 'Shrinkage and creep properties of high early strength structural lightweight concrete', *Proceedings of the Institution of Civil Engineers, Part 2*, **55** (Sep. 1973), 635–46.
89. R. N. Swamy, 'Shrinkage characteristics of ultra-rapid-hardening cement', *Indian Concrete Journal*, **48**(4) (Apr. 1974), 127–31.
90. A. K. Bandyopadhyay and R. N. Swamy, 'Durability of steel embedded in structural lightweight concrete', *Materials and Structures, Research and Testing*, **8**(45) (May–Jun. 1975), 203–10.
91. R. N. Swamy and A. K. Bandyopadhyay, 'The elastic properties of structural lightweight concrete', *Proceedings of the Institution of Civil Engineers, Part 2*, **59** (Sep. 1975), 381–94.
92. R. N. Swamy, A. B. Ibrahim and K. L. Anand, 'The strength and deformation characteristics of high early strength structural concrete', *Materials and Structures, Research and Testing*, **8**(48) (Nov.–Dec. 1975), 413–23.
93. R. N. Swamy and A. B. Ibrahim, 'Flexural behaviour of reinforced and prestressed Solite structural lightweight concrete beams', *Building Science*, **10**(1) (1975), 43–56.
94. R. N. Swamy and A. K. Bandyopadhyay, 'Shear behaviour of structural lightweight concrete T-beams without web reinforcement', *Proceedings of the Institution of Civil Engineers, Part 2*, **67** (Jun. 1979), 341–54.
95. J. Bobrowski and B. K. Bardhan-Roy, 'Structural assessment of lightweight aggregate concrete', *Concrete*, **5**(7) (Jul. 1971), 229–34.
96. *Lightweight Aggregate Concrete—Technology and World Applications*' (Editor: G. Bologna), Cembureau, Paris (1974), 210–13

97. Ref. 26, 50–1, 54–5.
98. Ref. 5, 533.
99. A. Short, 'Lightweight aggregate concrete: Part 1', *Concrete*, **8**(7) (Jul. 1974), 47–8.
100. Ref. 5, 261–4.
101. Ref. 25, 153–8.
102. Ref. 27, 68.
103. ibid., 71.
104. Ref. 25, 161.
105. F. C. Wilson, 'Concrete mix proportioning for pumping—some influences of material properties—a graphical approach to optimum gradation', *Proportioning Concrete Mixes*, Publication SP-46, American Concrete Institute, Detroit (1974), 25–45.
106. Ref. 27, 51.
107. F. D. Lydon, 'Some criteria for the choice and use of lightweight concrete', *Build International*, **6** (1973), 321–38.
108. F. D. Lydon, 'Artificial aggregates for concrete', *Concrete*, **9**(9) (Sep. 1975), 49–52.
109. R. N. Swamy, 'Prestressed lightweight concrete', *Developments in Prestressed Concrete—1* (Editor: F. Sawko), Applied Science Publishers Ltd, London (1978), Chapter 5.
110. 'Lightweight concrete pumped', *Concrete*, **7**(2) (Feb. 1973), 12.
111. R. Landgren, 'Water-vapor adsorption–desorption characteristics of selected lightweight concrete aggregates', *ASTM Proceedings*, **64** (1964), 830–45.
112. R. Landgren, 'Determining the water absorption of coarse lightweight aggregates for concrete', *ASTM Proceedings*, **64** (1964), 846–65.
113. F. D. Lydon, 'Some points on water absorption by lightweight aggregates', *Proceedings of the First International Congress on Lightweight Concrete, London*, May 1968, The Concrete Society, London (1970), Volume 2 (discussion), 12–18.
114. Ref. 26, 14–16.
115. Ref. 25, 69.
116. Ref. 27, 59.
117. ibid., 58–9.
118. G. Verbeck and L. E. Copeland, 'Some physical and chemical aspects of high pressure steam curing', *Menzel Symposium on High Pressure Steam Curing*, Publication SP-32, American Concrete Institute, Detroit (1972), 1–13.
119. Ref. 5, 276–80.
120. ibid., 530–1.
121. Ref. 26, 15–17.
122. Ref. 27, 88.
123. J. J. Williamson and R. K. Mackenzie, 'Sound insulation of lightweight concrete', *Build International*, **4**(4) (Jul.–Aug. 1971), 244–52.

4

The Use of Slate Waste in Road Construction

The use of slate waste as a road construction material has been touched upon, both directly and indirectly, in previous chapters. Chapter 2 included some discussion on the use of slate powder as a filler in bituminous materials specifically intended for road construction, and Chapter 3 considered, in general terms, the use of slate in various forms for making Portland cement concrete. However, there are other possible applications in road making.

There have been a few publications covering this field and these have indicated that there is some potential, but the actual usage of slate waste has been generally rather limited apart from some instances where advantage has been taken of local circumstances. A recent OECD (Organisation for Economic Co-operation and Development) report[1] noted that research and development have indicated the technical feasibility of using slate waste as an aggregate in bituminous concrete surface course, bituminous concrete binder course, and base and sub-base construction as well as for embankment fill and improved sub-grades. The report reviewed the use of waste materials and by-products for road construction by twelve OECD members and Table 4.1 indicates the broad usage of slate waste in these countries; unfortunately quantitative data is not available at present.

It is, in fact, understood that small quantities of slate waste are used for road foundation work in some industrial regions of France.[2] Furthermore, although the overall relative consumption of slate waste for road construction may be very small in the United States, one slate-producing company has found a good market for this application and the material is used by the Department of Highways and Transportation in Virginia.[3,4] In Finland,[5] slate quarries are located in the remote areas where little road construction activity exists and more suitable materials can usually be obtained readily. Furthermore, the extra cost of crushing the slate waste makes it even less competitive, and it is not used in bituminous mixes because of its particle shape. Nevertheless, it has been used in some road construction applications in Finland, but only in local circumstances and for secondary roads. Although not included in the OECD report, Norway[6] has a very considerable slate waste problem, and the industry there is working hard to find applications for this material; however, slate has not been recommended for road construction, mainly because of the poor particle shape, and potential uses are understood to be confined to filler applications and to bituminous roofing materials.

Table 4.1. The production of slate waste and usage for road construction in twelve OECD countries (from ref. 1).

Denmark Netherlands Switzerland	Not produced
Canada France Germany United States	Produced but not used for road construction
Belgium Finland Italy United Kingdom	Produced but only small quantities used for road construction
Spain	Produced and used for road construction

Much of the generally available published information about the utilization of slate waste in road construction has come from the United States and the United Kingdom, and the remainder of this chapter will be generally confined to work carried out in those countries.

A report by Marek, Herrin, Kesler and Barenberg, published in the United States by the Highway Research Board in 1972,[7] pointed out that highway construction and maintenance in the United States was consuming sand, gravel and crushed stone supplies at a steadily increasing rate. The authors noted that, although the nation's overall supply of aggregate was sufficient to meet any foreseeable demand, and plant capacity was also adequate, the unequal distribution of sources had created local and regional shortages that were critical. Furthermore, it was anticipated that the areas of shortage would broaden and that the shortages would become more intense. The work described in this report was undertaken principally to provide a systematic exploration of the processes (mainly technological) that could at that time, or through future development, produce additional and perhaps improved aggregates where conventional aggregate production was insufficient. The report pointed out that the United States was not alone in facing the problem of a diminishing aggregate supply; other countries, including the United Kingdom, were experiencing a similar problem.

The problem is particularly serious for highway construction and maintenance because such a high proportion of the total materials required consists of aggregates;[7] although Portland cement concrete may require (on average) about 75% aggregate, base courses and sub-bases contain 95–100%, and bituminous base, binder and surface courses may contain more than 90%. The report discussed a number of ways in which the problem of diminishing aggregate supplies can be approached. Some of these could make good use of waste materials (although it was noted that alterations of existing practices might be required for the construction of roads with materials other than conventional aggregates). In the case of slate, this might involve the manufacture of synthetic aggregate, either for use as a supplement or as a replacement for conventional aggregates, or simply the use of crushed material as discussed later in this chapter.

The report suggested various areas of research that appeared to have potential for yielding significant practical results. One of these specifically noted naturally occurring shales, clays and slates and divided them into (a) those materials suitable for the production of synthetic aggregates, (b) materials suitable for the production of brick, but not for synthetic aggregates, and (c) materials unsuitable for brick or aggregate production but suitable for upgrading by the use of additives. Another proposal suggested an investigation into hydrothermal hardening of pelletized calcium hydrosilicate bonded synthetic aggregates for highway use. This proposal pointed out that hydrothermally synthesized aggregates, made from materials containing appreciable proportions of silica or silica and alumina, may be economically viable if mixtures containing as little as 5% of Portland cement or lime are formed into highly compacted water-bound pellets. The heating and cooling stages of the autoclaving cycle may possibly be more rapid than in the case of the commercial autoclave curing of more conventional calcium hydrosilicate products, such as those described in the next chapter, because of the relatively small size of the pellets; furthermore, it is possible that reduced periods at maximum temperature may be found to give products with adequate properties. The report suggested that a single large commercial autoclave (3·4 m by 45·7 m in size) might process, in six cycles, up to about 1150 m^3 of pellets per day.

In a prepared paper, contained within this report, Willson[8] confirmed this proposal by suggesting that there is a potentially important area for research and development in the field of lightweight and dense processed aggregates. He pointed out that clays, shales and slates are widespread in the United States and, although many such materials do not expand well on heating to form lightweight aggregate, the sintering process (mentioned in Chapter 3) does not necessarily require a naturally bloating raw material. The addition of crude oil, and possibly other materials, may be used to improve or control bloating. He noted that this type of processed aggregate had already established its ability to produce satisfactory Portland cement concrete or bituminous concrete for most types of highway construction. In another paper, in the same report, Newlon[9] referred to earlier work[10] showing that heat-treated slate has possible uses as synthetic aggregate in sub-base, base, rigid and flexible pavement, as well as in reinforced (including prestressed) concrete applications. He also referred to the research by McGhee and Ozol[11] (described in some detail in Chapter 3) showing that untreated crushed slate can possess important properties (such as resistance to freezing and thawing, strength and chemical stability) which are satisfactory; furthermore, acceptable concrete could be made economically with appropriate attention to mix design. However, the particle shape of crushed slate would normally discourage its use where better aggregates are available.

A later report, by Miller and Collins,[12] was published in 1976 by the Transportation Research Board in the United States; this presented the results of a comprehensive survey of the technical and economic potential for making use of waste materials as alternative aggregates in highway applications. Of the initial group of 34 waste materials that were investigated, 31 were considered to show technical promise; and slate waste was among those that were regarded as deserving consideration for further development as aggregates.

For the initial screening of these materials, four minimum criteria were established:

(i) the annual quantity of waste material available at any one location must be at least about 45 000 tonnes—if annual production is less than that amount, then a quantity of at least about 450 000 tonnes must have accumulated;

(ii) the sources of waste materials must be located within a reasonable distance of potential market areas—the maximum hauling distances selected were 40 miles (64 km) by truck, 100 miles (161 km) by rail and 300 miles (483 km) by barge;

(iii) the material must not be highly toxic, especially after having been completely processed;
(iv) the material must not be soluble in water—it must not dissolve or leach out potentially harmful substances.

The scheme used for the technical evaluation of the waste materials was complex, and a detailed description of it is beyond the scope of this discussion, but due consideration was given to their properties in relation to the different types of highway applications contemplated. General as well as specific physical, mechanical and chemical properties were considered, together with relevant thermal and optical properties (such as their coefficient of thermal expansion and their reflectivity). Wastes having maximum potential for general aggregate use, for example, should have adequate soundness, hardness, gradation and particle shape, and resistance to chemical and physical deterioration; furthermore, they should require a minimum amount of processing (such as crushing, grading and blending) before they are used.

The waste materials were classified into four groups depending on their potential, in particular highway applications, as determined by the technical evaluation. There was no application for which slate waste was included among the materials having the highest technical potential for use. Nevertheless, it was grouped among the second best class of materials for use as aggregate in stone base applications. It was among the third class of materials for general use as highway aggregate, for use as aggregate in bituminous mixtures and for use as aggregate in Portland cement concrete; the report noted that, for a given application, the third group of materials would generally only be recommended for use in local isolated cases. The fourth group of materials, which did not include slate for any application, contained those wastes that showed little or no promise for use as aggregate.

Processing costs were estimated for some of the waste materials, but not for slate, and transportation costs were also discussed; however, it was noted that, depending on specific circumstances, costs of this kind are subject to wide variations.

So, as a guide to the overall economic feasibility, each waste material was evaluated in terms of various factors that were selected to reflect its economic potential for development as an aggregate. In general terms, these were as follows: first, the location and available quantity of each material in relation to potential market areas, aggregate deficiencies, and access to transport facilities; secondly, the actual use, or potential applications for use, of the material as an aggregate in highways; thirdly, the possible resource value of the material due to other potential or developed uses, its heat content where applicable (e.g. coal mining waste), and the extent of processing required in order to produce a useable product; and, finally, ecological and social considerations including such factors as the presence of a nuisance or ecological blight, governmental interest in development of uses for a specific waste material, backing from special interest groups for certain wastes, and conservation efforts possibly directed at specific waste materials.

According to the report, the various factors found to influence the economic feasibility of developing slate waste as a highway aggregate in the United States can be generally summarized as follows:

(i) *Location and quantity:* although the quantities involved are not large (relative to sources of some other waste materials), they are located near potential markets, which are close to areas with aggregate shortages, and where there is a growing demand for aggregates.

(ii) *Application in highways:* slate waste is currently acceptable as a highway material and its principal use is (or would be) for highway construction; it possesses properties to recommend its use in highways, and applications which are technically feasible include stone base construction and Portland cement concrete.

(iii) *Resource value:* slate waste exists in some quantity as an inert and combinable material; crushing and sizing are the only required processing steps (unless it is to be used for the production of a high quality product such as lightweight aggregate) and a high percentage of the available waste material is usable after processing.

(iv) *Ecological and social considerations:* there seem to be no important factors of this kind related to the development of slate waste aggregate in the United States.

On the basis of the economic evaluation, the materials were again divided into four classes, depending on their overall economic feasibility; slate waste was listed among the materials in the second class, which includes those demonstrating some favourable aspects for utilization as highway aggregates. The report did not reveal any particular environmental factors which would either tend to preclude or encourage the recycling, processing or actual use of slate waste in road construction; nevertheless, it was listed as one of the materials recommended for development as aggregate on the basis of environmental considerations.

The report provided an overall evaluation of the waste materials which was based on the results of the separate technical, economic and environmental evaluations; and this led to their being grouped into four classes, from I to IV. Class I refers to those materials that show the most potential for use as highway aggregates and includes blast furnace slag and fly ash, which have both evolved as acceptable highway materials over a period of many years. Nearly all the Class I materials have received some measure of acceptability for highway use. All of them possess the desirable properties of aggregates and are capable of treatment; and, generally, they are located in or near metropolitan areas and occur in sufficient quantity to provide a continuous source of supply for highway construction. Class II, which includes slate waste, covers those materials that deserve consideration for further development as aggregates but which, at the time of the report, did not appear to have as high a potential as the Class I materials. Class II materials do not always possess so many favourable properties, and they generally require a greater amount of processing to be rendered suitable for use as an aggregate. Although most of them exist in significantly large quantities, many of them are not located within the immediate proximity of potential market areas. Several Class II materials, including slate waste, have been used to a limited extent in highways. Class III materials are those which do not show great promise for use as a highway aggregate for a variety of reasons—and Class IV are those not recommended for further consideration.

The report referred to the strength, stability and durability of crushed slate, and the undesirable flat, elongated particle shape. It suggested that the actual use of slate waste in the United States had been very small and briefly discussed the usage in Virginia, noting that, in 1972, a total of about 27 000 tonnes had been used there by the Department of Highways and Transportation.

In Virginia, slate is currently being used for incidental concrete construction, for base material, for stabilization of secondary gravel roads, and as the coarse aggregate in hot mixed bituminous concrete bases and surfaces.[4] While many thousands of tonnes are used annually as base material, the bituminous concrete application is still in the experimental stage as the long-term skid resistance of slate mixtures is being evaluated. According to one slate producer in Virginia,[3] although the particle shape of the crushed slate waste is somewhat flat or elongated, this has presented no more than token problems and that particular material has found a good market as a construction aggregate. Its relative hardness, compared with other slates in the United States, is important and many softer slates would probably fail to meet the necessary specifications (it has a hardness number of 6 on Mohs' scale, a magnesium sulphate soundness of 2–2·4, and a Los Angeles abrasion value of 18–20).

In a Transport and Road Research Laboratory report, published in the United Kingdom in 1974, Sherwood[13] pointed out that it is in the national interest to make use of lower grade and waste materials as alternatives to naturally occurring aggregates as this conserves the supplies of good quality aggregates and assists in problems arising from the disposal of unwanted materials. In addition, economic considerations make it likely that the cost of natural aggregates will increase in those areas where they are already in short supply whereas low-grade and waste materials are by their very nature available at low cost, at least at source.

Sherwood showed that, in the United Kingdom, the total quantity of waste materials available far exceeds the amounts of aggregates and fill material required for roadworks, assuming that all the wastes are suitable to act as substitutes for all the aggregates used in the pavement layers and for the materials used as imported fill. He pointed out that, of course, the scope for substitution is severely limited by the unsuitability of the wastes or by the non-availability within an economic haulage distance. Although the use of waste materials in road making can play a part in reducing the accumulation of waste tips, he suggested that this is a relatively minor role; the major reasons for encouraging utilization are the conservation of good quality materials that can be put to better use elsewhere at a future date and the avoidance of disturbing the landscape (and possible consequent dereliction) by the excavation of borrow pits to obtain road making materials. Thus there are considerable environmental advantages to be gained from the use of wastes.

In his report, Sherwood considered the extent to which the road construction industry can use low grade and waste materials and discussed the types and quantities of materials available, their various applications in road construction, and the tests needed to ensure that they meet the required specifications. He suggested that, in the United Kingdom, slate waste has little potential for road making except as a fill material, and that it is unlikely to find much use in this application because it is generally situated in remote areas.

Subsequent reports, in the same series, considered waste materials which occur in large quantities and which have high potential outlets in road construction. Apart from materials of this kind, which include PFA and colliery shale, there are those which are inherently unsuitable, and there is also a third group consisting of materials produced in fairly large amounts but which have found less application in road making for various reasons. Sherwood, Tubey and Roe considered these materials, and included slate waste among them, in a report published in 1977;[14] they attributed the general lack of utilization of these largely to difficulties in making them available at a given time in sufficient quantities at a reasonable price, but they noted that other factors may also be involved. As far as slate is concerned, they stressed the remote locations of the very large available quantity as the limitation to its value as a road making material; were it not for this factor, they suggested that more research into its properties in this application would be justified. However, they noted that it has been used for road works in areas where it is produced, and their report shows an embankment made from slate waste in a section of road improvement on Anglesey in North Wales. The flakey nature of the slate waste particles has caused problems in compaction, and grid rollers have been found to be useful since these break the elongated pieces of slate into shorter pieces; using this technique, slate waste has been successfully employed for sub-base construction in road works in North Wales.[14,15]

In this particular case,[15] very variable ground and ground water conditions imposed considerable difficulties in designing the road structure for the improvement of a two-lane road over a length of approximately 3 km. The design called for an overall road construction thickness of 800 mm (maximum) comprising 520 mm of sub-base, 190 mm of roadbase and 90 mm of blacktop; this construction was required to give an even surface in the very variable conditions encountered.

The material available had no cohesive fine fraction but did contain a proportion of large 'dendritic' pieces. Crushing and/or screening were investigated as a means of eliminating material larger than 75 mm, but it was found that this was an uncertain as well as an uneconomic process; the slate pieces could be 500 mm long though still less than 75 mm across. The most suitable material was found to be waste from machine dressing, comprising material generally below 100 mm, from which the occasional large piece could be hand picked. Keeping the material at a low level in the road enabled the specification to be varied to accept a proportion of oversize pieces. It was agreed to use the material in the lower 350 mm of sub-base where the question of susceptibility to frost damage would not arise.

The material was put down in two layers, each rolled with twelve passes of a grid roller (see Fig. 4.1) and six passes of a vibrating roller; roll pressures were 0·73 kg/mm and 2·33 kg/mm respectively. The initial standard of rolling was determined by field sand replacement density tests and, from time to time in the course of the work, sand replacement tests were made as a check on the work.

As noted above, it was found that the action of the grid roller broke the longer needle-shaped pieces of slate into short pieces, thus improving compactibility; furthermore, the process of breaking did not cause the material to become clayey. Generally the fines in the materials seemed to be

Fig. 4.1. Slate waste used for sub-base construction (from ref. 15, by courtesy of Mears Construction Ltd).

in the sandy rather than in the clayey range; and little trouble was experienced with compaction, even in very wet weather.

(It should be noted that some slate samples from North Wales have recently been shown to perform satisfactorily in frost susceptibility tests carried out in accordance with TRRL Report LR 90.[16])

The OECD report,[1] noted earlier, was based on research and experience in twelve member countries including the United States and the United Kingdom; it therefore tends to give a more overall view of the use of waste materials. The report included an inventory of materials having potential for road construction; and it described established and possible applications as well as discussing strategies for using the materials taking into account the various technical, economic, energy and environmental aspects. It noted some of the incentives and inhibiting factors which tend to affect the utilization of wastes, and that increased constructional activity had led, in many areas, to a local depletion of the naturally available materials which are traditionally used by the industry. The large increase in aggregate production between the mid-1950s and 1975 was illustrated for several countries; and although there was a sudden reversal in the trends in demand following the energy crisis in 1973, it was anticipated that growth would continue (albeit at a much lower rate) and there would therefore be a continuing need for aggregates in urban areas. However, the enormous quantities of aggregates that had already been used meant that, in many urban areas, local materials were no longer available and the deficit had to be made good by the transport of materials from further afield. And, as noted from other sources, this report also made the point that utilization could relieve some of the pressure on natural resources as well as disposing of waste materials and, in some cases, would release land rendered or about to be rendered derelict by waste tips.

As with the earlier report published in the United States,[12] the general criteria for a material to be classified as usable in road construction were based on available quantities, transportation distance, toxicity and solubility in water. And, again, the materials were grouped into four classes, from I to IV, based on an overall evaluation. Class I included materials with maximum potential, possessing the best properties either in their naturally occurring, processed or combined form, or having a record of satisfactory performance. Class II materials require more extensive processing and/or have properties not as adequate as those in Class I. Class III materials show less promise than I or II and are recommended for use only in isolated situations. And Class IV materials show little or no promise for use in road construction applications.

Table 4.2 lists the Class I, II and III materials, grouped by order of merit within each class, but it should be noted that this is based on the overall experience of the twelve participating OECD member countries; obviously each individual country may have a somewhat different evaluation listing based on such factors as availability and local experience.

Slate waste lies over half way up Class II and therefore appears to have some significant potential within the participating countries. However, apart from some exceptions, there seems to be little evidence that the overall utilization of slate waste in road construction is matching its potential. It seems that the inhibiting factors are outweighing the incentives; and this aspect of the slate waste problem will be discussed further in Chapter 8.

Table 4.2. Overall waste material and by-product evaluation for road construction in twelve OECD countries (from ref. 1).

Class I	*Class II*	*Class III*
Blast furnace slags	Phosphate slag	Gold tailings
Fly ash	Copper slag	Copper tailings
Sulphur	Quarry waste	Lead–zinc tailings
Steel slag	Mine refuse	Coke breeze
Boiler slag	*SLATE WASTE*	Foundry sand
Bottom ash	Bark and sawdust	Ceramic and refractory wastes
Nickel slag	Fluorspar tailings	Resin
Demolition wastes	Tyres and rubber	Lignin
Colliery spoil	Zinc slag	Potassium salt
Oil shale residue	Lead slag	Mine waste
Taconite tailings	Mixed ash	Pyrite cinders
China clay sand	Incinerator residue	Waste glass and cullet
Iron ore tailings		Waste plastics
Pyrolysis residue		Cellulosic

REFERENCES

1. *Use of Waste Materials and By-products in Road Construction*, Organisation for Economic Co-operation and Development, Paris (1977).
2. R. Reverend (Director, Technical Department, Fédération des Ardoisières de France), private communication.
3. R. G. Cabell (President, Arvonia-Buckingham Slate Company, Inc.), private communication.
4. K. H. McGhee (Senior Research Scientist, Commonwealth of Virginia Highway and Transportation Research Council), private communication.
5. M. Eerola (Road and Traffic Laboratory, Technical Research Centre of Finland), private communication.
6. A. Grønhaug (Public Roads Administration, Oslo), private communication.
7. C. R. Marek, M. Herrin, C. E. Kesler and E. J. Barenberg, *Promising Replacements for Conventional Aggregates for Highway Use*, National Cooperative Highway Research Program Report 135, Highway Research Board, Washington, D.C. (1972).
8. C. Willson, *Technical Capabilities—Current and Future*, ibid., 18–21.
9. H. H. Newlon, Jr, *Replacement of Alternate Materials for Existing Aggregates*, ibid., 23–8.
10. F. F. Fondriest and M. J. Snyder, *Synthetic Aggregates for Highway Construction*, National Cooperative Highway Research Program Report 8, Highway Research Board, Washington, D.C. (1964).
11. K. H. McGhee and M. A. Ozol, *An Evaluation of a Slate Aggregate for Use in Concrete*, Virginia Highway Research Council (Nov. 1966).

12. R. H. Miller and R. J. Collins, *Waste Materials as Potential Replacements for Highway Aggregates*, National Cooperative Highway Research Program Report 166, Transportation Research Board, Washington, D.C. (1976).
13. P. T. Sherwood, *The Use of Waste and Low-grade Materials in Road Construction: 1. Guide to Materials Available*, Transport and Road Research Laboratory Report 647, Department of the Environment (1974).
14. P. T. Sherwood, L. W. Tubey and P. G. Roe, *The Use of Waste and Low-grade Materials in Road Construction: 7. Miscellaneous Wastes*, Transport and Road Research Laboratory Report 819, Department of the Environment and Department of Transport (1977).
15. P. Titman, 'Slated in North Wales', *Contact—The Journal of the Mears Group*, Garnett-Keeler Public Relations Ltd, Surbiton, Surrey (Spring 1976); also P. Titman (Planning and Research Manager, Mears Construction Ltd), private communication.
16. R. H. Boyle (Director, Penrhyn Quarries Ltd), private communication.

5

Autoclaved Products from Slate Waste

5.1. INTRODUCTION

Autoclaved calcium hydrosilicate products form an important group of constructional materials that provides a wide range of useful engineering properties with a variety of applications. Such products include sand/lime bricks, autoclaved aerated concrete blocks and reinforced units, and autoclave cured concretes containing dense or lightweight aggregates; there are also various asbestos-fibre reinforced materials in the form of pipe, and both corrugated and flat sheets, as well as lightweight insulation products.

The bonding materials present in these products consist of calcium hydrosilicates that are formed by autoclaving (i.e. high-pressure steam curing) mixtures of lime and silica. (Although silica in the form of sand, for example, is usually regarded as unreactive under normal conditions, it can behave as an acidic oxide at high temperatures and react with basic materials to form other compounds.) The use of different mixtures of raw materials and of different manufacturing methods provides a variety of useful combinations of mechanical properties, stability, durability, density and, for lightweight products, good thermal insulation properties. The source of lime is often Portland cement, or lime itself, or a combination of the two; and the siliceous component may be provided by materials such as pulverized fuel ash or natural sand.

Analysis of the gross chemical composition of most slates gives a value for the silica content which generally falls between about 45 and 65 %. This suggests that slate waste may have potential uses in the manufacture of autoclaved materials, and this chapter describes some of the research and development work that has been aimed at investigating the use of slate in making products of this kind.

Recent work by Watson, Grant and Jones[1] provides a convenient introduction. These workers have investigated some of the factors influencing the compressive strength of autoclaved materials made from mixtures of ordinary Portland cement and fine slate powder; 94 % of the slate powder which they used passed through a 0·075 mm sieve, 88 % passed 0·053 mm, and it was from the same source as the material for which the chemical analysis is given in Table 5.1.[2] The slate content of these mixtures was varied between 0 and 80 % and pastes were formed, by mixing with water, using initial water/total solids ratios between 0·35 and 0·425. After

Table 5.1. Chemical analysis of slate powder from the same source as that used for making experimental autoclaved slate powder/Portland cement compositions[1,15,16,23] (from ref. 2).

Silica	59·85 %
Alumina	21·50 %
Iron oxide	8·10 %
Titanium dioxide	0·55 %
Lime	1 73 %
Magnesia	2·40 %
Sodium oxide	0·03 %
Potassium oxide	0·15 %
Phosphorus pentoxide	0·15 %
Manganese oxide	0·19 %
Sulphate	1·32 %
Carbon dioxide	0·05 %
Loss on ignition	3·07 %
Not determined	0·91 %
	100·00 %

casting and the initial hardening of the pastes, the 'green' test specimens were demoulded and precured under water for varying periods, up to 28 days from casting, before they were autoclaved. The form of the autoclave cycle used in this work, and in other investigations by Watson and his co-workers which will be referred to later, was generally as indicated in Fig. 5.1; manual control was used and the shaded area shows the limits within which this could typically be achieved. The temperature was raised to 170 °C over a period of about 3–4 hours, and more slowly to about 200–205 °C over the next 4 hours; it was then rapidly lowered to 100 °C and finally allowed to fall to ambient temperature overnight.

The specimens were then removed from the autoclave and tested immediately. Figure 5.2 shows the effect of slate content (using a constant

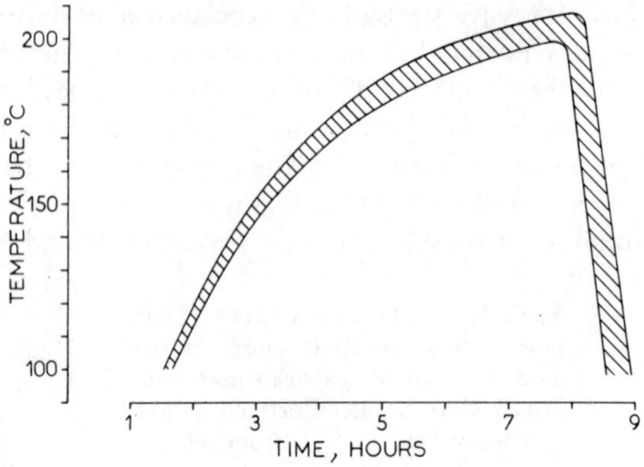

Fig. 5.1. The autoclave cycle used for the preparation of experimental autoclaved slate/cement compositions by Watson and his co-workers[1,15,16,23] *(from ref. 16).*

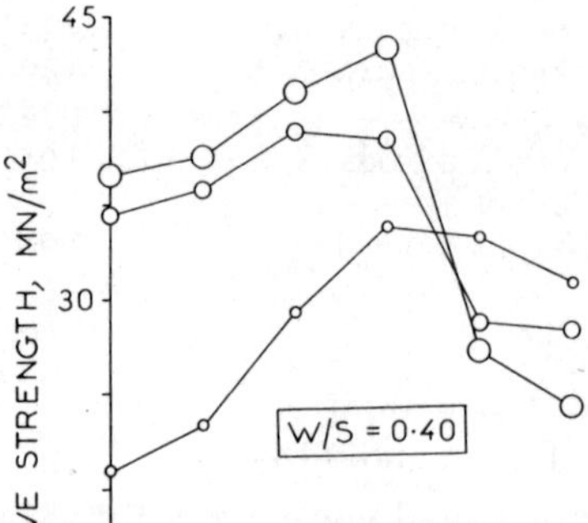

Fig. 5.2. The effect of slate content and precuring time on the compressive strength of autoclaved slate/cement mixes with a water/solids ratio of 0·40 (from ref. 1).

initial water/total solids ratio of 0·40) on the compressive strength of specimens that had been demoulded the day after casting and autoclaved either immediately (1 day) or after they had been precured under water until 7 or 28 days after casting. Maximum strength occurs at slate contents of about 30–45% and becomes greater as the precuring period before autoclaving is increased; up to 45% there is an overall increase in strength with precuring time, but at 60 and 75% there is a corresponding decrease.

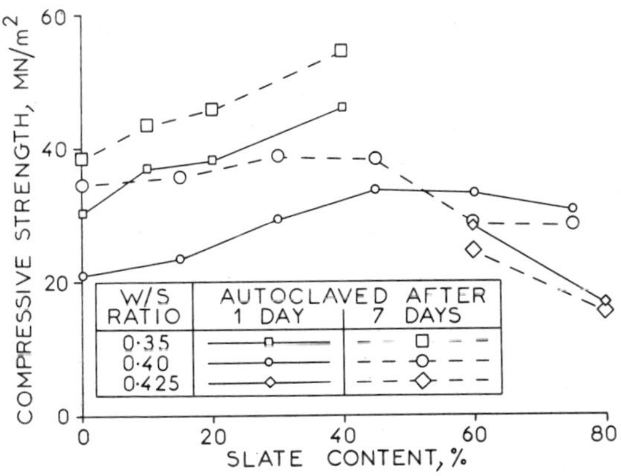

Fig. 5.3. The effect of water/solids ratio, slate content and precuring time on the compressive strength of autoclaved slate/cement mixes (from ref. 1).

Figure 5.3 shows the result of varying the water/total solids ratio between 0·35 and 0·425 with precuring times of 1 and 7 days. The overall effect is similar to that of varying the water/cement ratio in normal concrete; other factors being equal, an increase in water/total solids ratio leads to a decrease in compressive strength.

These results include some respectable compressive strengths, particularly for slate contents in the region of 30–45%; and this peak in strength is similar to those observed in the well-known investigations, using other siliceous materials, that were carried out by Menzel in the early 1930s.[3] The underlying chemical reactions leading to the formation of the various calcium hydrosilicates under autoclaving conditions are complex and further discussion of these is outside the scope of this book, but the more fundamental details are reviewed elsewhere.[4–6] Furthermore, the reasons for the varying strength obtained under different conditions are also complex and far from properly understood. However, it is interesting to note that Verbeck and Copeland[7] point out that an overall molar lime/silica

(CaO/SiO_2) ratio of about 1, in mixtures of this kind, gives autoclaved reaction products with relatively high strength; and, in fact, the slate content of 45% in Watson, Grant and Jones' work corresponds to a bulk lime/silica ratio of 1·0.

However, slate is a complex, variable material and analysis generally reveals an alumina content between 11 and 25% (21·5% in the case of the slate used by Watson, Grant and Jones) together with significant quantities of iron compounds as well as other components. Scientific studies have suggested that the addition of Al^{3+}, in suitably reactive form, tends to increase the 'crystallinity' of the reaction products formed by autoclaving, and there is indirect evidence in the fact that it also reduces drying shrinkage.[8] For instance, Noorlander[9] has produced results which show that a marked reduction in the shrinkage of calcium hydrosilicate bricks was achieved by adding aluminous materials to the mixtures from which they were made. Of course, drying shrinkage is an important property in constructional applications and will be further discussed later in this chapter.

Over the course of many years, there have been a number of references to the possibility of making autoclaved products from slate and from similar materials.[10–12] The Mineral Industries Experiment Station programme[13] examined methods of making bricks and aerated acoustical tiling materials from slate. In 1954, the Building Research Station, in the United Kingdom, published a report[14] of an investigation into making autoclaved bricks from slate waste produced in Scotland; subsequently, similar work was carried out using slate from North Wales.[2] And, within the last few years in the United Kingdom, there have been investigations into the use of slate powder for making autoclaved aerated concrete.[15–17]

The following discussion is divided into two parts; the first considers the production of slate/lime bricks and the second discusses the work that has been carried out with aerated materials.

5.2. SLATE/LIME BRICKS

Work carried out at the Mineral Industries Experiment Station[13] suggested the possibility of producing autoclaved slate/lime bricks, similar to sand/lime bricks. The results indicated that crushed slate waste passing a 20 mesh sieve, mixed with about 10% hydrated lime, could be pressed and autoclaved to produce bricks and other building products.

Initial laboratory experiments involved small test tiles autoclaved at a

Table 5.2. Modulus of rupture, compressive strength and porosity of various slate/lime brick compositions compared with a sand/lime composition (from ref. 13).

	Modulus of rupture (MN/m^2)	Compressive strength (MN/m^2)	Porosity (%)
(Sand and slate of same grading) Sand/lime	4·82	40·40	21·8
(Sand and slate of same grading) Slate/lime	3·29	24·48	33·2
Coarse slate (fines removed)	3·53	23·44	29·6
Run of mill slate, 7·5% lime	5·98	42·13	20·7
Run of mill slate, 10·0% lime	7·49	42·13	24·9
Run of mill slate, 12·5% lime	6·81	30·06	20·1

pressure of 1·03 MN/m^2 for about 7 hours. Modulus of rupture values ranged between 7·70 and 9·18 MN/m^2 (with porosities of 24·4–34·1%) depending on the grading, and corresponding values of 2·56–6·76 MN/m^2 and 28·6–33·2% were obtained with sand/lime compositions.

Full-sized bricks (64 × 114 × 229 mm) were then made on a commercial press and autoclaved; the modulus of rupture and compressive strength values of these materials are shown in Table 5.2. (The best compressive strength values of 42·13 MN/m^2 compare well with a quoted value of 29·65 MN/m^2 for common brick, although a value of 55·85 MN/m^2 was given for hard-fired shale face brick.)

It was noted that raw or calcined slate could be used, and expanded slate was also tried. The expanded material was crushed to below 8 mesh (with the formation of a large proportion of fines) and mixed with different amounts of lime; the pressed specimens were autoclaved for 7 hours at 1·21 MN/m^2 and some properties are shown in Table 5.3. The report suggested that, whereas the optimum lime content for the raw slate seemed to be about 10%, the expanded slate needed more.

Table 5.3. Some properties of autoclaved slate/lime materials based on raw and expanded slate (from ref. 13).

Composition (%)		Modulus of rupture (MN/m^2)	Absorption (%)	Porosity (%)	Density (kg/m^3)
Slate	Lime				
90% raw	10	8·27	14·5	30·0	2 214
90% expanded	10	3·34	35·0	44·1	1 179
80% expanded	20	5·76	33·9	44·6	1 295

The Building Research Station investigation,[14] which used a slate waste from Scotland, examined the effects of lime and water contents, slate grading, and autoclaving conditions.

The first stage was aimed at determining the optimum lime content. This involved preparing 38 mm diameter test cylinders from crushed slate, screened to pass a 4·76 mm sieve, mixed with hydrated lime. The lime content was varied from 3 to 7% CaO (based on the slate) and water was added (2% based on the slate, plus 10% on the hydrated lime) before pressing at 30·89 MN/m^2. The specimens were autoclaved for 4 hours at 1·10 MN/m^2, and then stored in water for 24 hours before their compressive strength was determined; the mean of triplicate determinations ranged between 8·76 and 13·31 MN/m^2 with increasing lime content, and the results indicated that 4% CaO provided an acceptable compromise between strength of the material and economy in the process. (There was considerable scatter in the results, which was attributed to the different positions of the specimens in the autoclave—and later work avoided placing the specimens too close to the water at the bottom.)

The next stage involved varying the water content of specimens containing 4% CaO (5·2% hydrated lime); the proportion of water was varied between 2·5 and 6·5% (0·5% based on the lime content plus 2–6% on the slate). A water content of 4·5% gave the highest compressive strength without exuding moisture during the pressing operation. Some difficulties were experienced with disruption of specimens during autoclaving, but these were overcome by drying them beforehand.

The preceding work had been carried out with crushed slate passing through a 4·76 mm sieve. The next stage compared the results of using material passing 7, 18 and 52 mesh sieves; corresponding compressive strengths of 18·48, 13·86 and 11·65 MN/m^2 respectively were obtained using lime and water contents of 4% and autoclaving at 1·10 MN/m^2 for 4 hours. Thus slate passing a 7 mesh sieve gave the highest strength (including the results obtained with the material passing 4·76 mm) although less between-specimen scatter was obtained with the finer powders.

The next part of the work involved the production of four 152 × 76 × 51 mm bricks using crushed slate passing a No. 7 sieve, 5·2% hydrated lime (4% CaO) and 4·2% water. A small machine press was used, giving a moulding pressure most probably below 7·72 MN/m^2, and after autoclaving at 1·10 MN/m^2 for 4 hours, the bricks had an average compressive strength of 20·69 MN/m^2. (It was noted that, if the previously used moulding pressure of 30·89 MN/m^2 could have been used, the strength would have been higher.)

A larger batch of bricks was then prepared for weathering tests; the processing conditions were similar to those used before, although the grading of the slate was slightly inferior, and the bricks had an average compressive strength of 18·41 MN/m^2 after 24 hours' storage under water. A pier was constructed from these bricks, at the Building Research Station, in order to investigate their weathering characteristics and six bricks were laid in the open, three in a shallow tray. The results can be seen in Figs. 5.4 and 5.5; judging from their appearance, the relatively well protected bricks in the pier seem to have weathered fairly well, even after nearly 14 years, but the single bricks exposed in the tray suffered severe damage after only a year.

Finally, small test cylinders were employed to determine the optimum autoclaving conditions using the same material that was graded for the large batch of bricks. After pressing at 30·89 MN/m^2, the cylinders were autoclaved for 4 hours at 1·10, 0·90 and 0·69 MN/m^2; corresponding average compressive strengths of 17·17, 17·86 and 14·27 MN/m^2 were obtained—and this suggested an optimum pressure of 0·90 MN/m^2. Furthermore, autoclaving at 0·69 MN/m^2 for 8 hours gave an average strength of 16·62 MN/m^2, almost equal to that obtained at 0·90 MN/m^2 for 4 hours.

The bulk density of the bricks produced in this work was found to be 2162 kg/m^3. And their drying shrinkage and reversible moisture movement values were found to be 0·046 and 0·041 % respectively. Unfortunately the drying shrinkage was nearly twice the value of 0·025 % that was required for the type of brick that would have taken advantage of the high compressive strength of the slate/lime composition. It was concluded that further work was needed to reduce this drying shrinkage.

Some years later, a similar investigation was carried out using slate from North Wales.[2] Tests were carried out with crushed slate passing 4·76 mm, 8 mesh, 16 mesh and 25 mesh sieves; bricks could be made from all these gradings but the most satisfactory were obtained from the 25 mesh material. Test specimens were produced in the form of small cylinders and prisms; mixtures of 90 % slate and 10 % hydrated lime were used and 5 % water was added. After pressing, the specimens were autoclaved at 1·21 MN/m^2 for 8 hours. The addition of silica to the mixture, as fine sand or silica flour, had no beneficial effect and this indicated that the slate contained sufficient reactive silica to provide an adequate bond.

Compressive strengths ranged from 30·34 to 45·51 MN/m^2; and it was noted that it should be possible to guarantee a value of 34·48 MN/m^2 which was the requirement for the highest quality calcium silicate bricks at that time.

For the best experimental bricks, made from the 25 mesh material,

(a) October 1954 (when built).

(b) October 1955.

(c) July 1968.

Fig. 5.4. A pier, built from autoclaved slate/lime bricks, undergoing weathering tests at the Building Research Establishment exposure site. (Crown copyright photographs reproduced by courtesy of the Director of the Building Research Establishment.)

(a) October 1954
(when placed).

(b) October 1955.

Fig. 5.5. Autoclaved slate/lime bricks undergoing weathering tests at the Building Research Establishment exposure site. (Crown copyright photographs reproduced by courtesy of the Director of the Building Research Establishment.)

drying shrinkage values of 0·040–0·043% were obtained, but values as low as 0·02% had been observed for materials using the coarser fractions of slate (although these latter materials tended to fail the frost resistance test); the British Standard requirement for the highest quality calcium silicate bricks was 0·025%. (Although the best specimens gave relatively high drying shrinkage values, that were nevertheless markedly lower than for some other types of bricks, it was noted that this would not necessarily be a serious disadvantage; for instance, the selection of the appropriate spacing between movement joints should avoid problems in large restrained walls.)

The bricks had a density of 2082 kg/m^3, which was markedly higher than that for clay and sand/lime bricks. It was noted that the higher density would be advantageous from the point of view of sound insulation; and, if lighter bricks were required, frogs or holes could be provided.

Water absorption tests gave values of 10–12 ml/100 g, which was within the range required for sand/lime bricks. And the best specimens performed well in frost resistance tests, indicating that they could have been used under extreme exposure conditions. A somewhat severe German standard frost resistance test was used which involved 25 freeze–thaw cycles. In two series of tests, the best bricks showed little or no visible damage afterwards as well as retaining 99% and 90–95% of their compressive strength (although bricks made with the coarser fractions failed the test). Furthermore, there was no evidence of any tendency towards efflorescence. The bricks had an attractive appearance, with smooth surfaces and sharp arrises; depending on the slate used, they were either a shade of grey that was popular for facing at that time, or a green colour that was not otherwise commercially available.

In view of the aluminium content of slate, the high drying shrinkage values that have been found for slate/lime bricks are perhaps rather surprising. Furthermore, the effect of the factors, and their interactions, which influence frost resistance seem to require further investigation. If any future development work on slate/lime bricks is contemplated, it would seem sensible to begin by examining these two aspects of their behaviour in more detail.

5.3. AERATED MATERIALS

The possibility of using slate waste for making aerated autoclaved materials has also been considered. Such materials, of course, have good thermal insulation properties and some have been regarded as having useful acoustical properties.

The Mineral Industries Experiment Station programme[13] included an investigation into making an autoclaved aerated acoustical tiling material. Initial trials began with a mixture of slate (that had been calcined at 1000 °C), hydrated lime, aluminium powder and water. (The aluminium powder reacts with the lime to generate hydrogen gas bubbles which give the material a cellular or 'aerated' structure.) The first part of the work involved

Table 5.4. The effect of calcining, autoclaving, aluminium content and water content on the modulus of rupture of experimental acoustical tiling materials based on equal quantities of slate and lime (from ref. 13).

Mix composition			*Modulus of rupture* (MN/m^2)	
Slate	*Al* (%)	*Water* (%)	*Not autoclaved*	*Autoclaved*
Raw	0·050	50	0·31	0·66
	0·075	50	0·24	0·52
	0·100	50	0·21	0·55
	0·125	50		0·55
Calcined	0·050	58	—	1·07
	0·075	55	0·17	1·24
	0·100	58	0·21	1·14
	0·125	72	—	0·69

determining the lime content that was required to provide sufficient 'green' strength so that the material could be handled before autoclaving; on the basis of this, equal quantities of slate and lime were used for the subsequent stages of the investigation. The effect of the water content of the mix was examined and it was established that, the more water used, the less dense and weaker the material became. The quantity of water used was generally dictated by the plasticity of the mix. Sufficient workability was needed for moulding but it was noted that the mix should be neither too wet nor too dry in order to optimize the expansion process; if it was too wet, it became 'frothy' with excessively large bubbles and, if too dry, the cake cracked and the expansion gas escaped.

Table 5.4 shows the effect of aluminium content, as well as that of calcining the slate and of autoclaving, on the modulus of rupture of materials based on equal quantities of lime and slate. The table indicates that calcining and autoclaving had a markedly beneficial effect on the modulus of rupture, and it should be noted that the calcined slate materials used higher water contents than the others. Furthermore, on the basis of these results, an aluminium content of 0·1 % was selected for further work.

Table 5.5. The effect of slate treatment, water content and autoclaving on modulus of rupture and tensile strength of experimental acoustical tiling materials based on equal quantities of slate and lime with 0·1 % aluminium (from ref. 13).

Slate	*Water (%)*	*Modulus of rupture (MN/m^2)*		*Tensile strength (MN/m^2)*	
		Not autoclaved	*Autoclaved*	*Not autoclaved*	*Autoclaved*
Raw	60·0	0·07	0·19	0·28	0·45
Calcined	48·3	0·32	2·61	0·21	0·90
Raw	43·3	0·39	1·97	0·21	1·00
Calcined	43·3	0·16	0·24	0·21	0·83
Expanded	43·3	0·68	1·11	0·24	0·86

(In an attempt to increase the plasticity of the mixes, and reduce the quantities of aluminium and water used, 5 % bentonite was added to the slate/lime blends but no improvement was observed.)

Table 5.5 shows the modulus of rupture and tensile strength of non-autoclaved and autoclaved materials prepared with equal quantities of lime and slate (using raw, calcined or expanded slate) and 0·1 % aluminium powder; autoclaving was carried out at 1·03 MN/m^2 for 7 hours. It is interesting to note that, in this series of materials, raw slate gave higher strengths (apart from non-autoclaved tensile strength) than the calcined slate at a water content of 43·3 %; furthermore, for the calcined slate materials, a water content of 48·3 % gave higher strengths than those obtained at 43·3 % (again, except for the non-autoclaved tensile strength).

A slate-based composition was compared with several commercial materials and the results of one of these comparisons are shown in Table 5.6. The experimental material contained the basic blend of equal

Table 5.6. The modulus of rupture and relative air permeabilities (compared with the slate sample at 30 mm water pressure) of an experimental slate-based acoustical tiling composition and a commercial material (from ref. 13).

Sample	*Modulus of rupture (MN/m^2)*		*Relative air permeability at pressures (mm of water) of:*		
			30	*183*	*782*
Slate[a]	0·24	0·21[b]	1·00	0·68	0·43
Commercial	0·47	0·12[b]	1·01	0·72	0·45

[a] Equal quantities of slate and lime, 0·3 % aluminium and 60 % water.
[b] After drying overnight at 110 °C.

quantities of slate and lime, with 0·3% aluminium powder and 60% water; its porosity and water absorption values were considerably below those of the commercial material (64% and 57% of the respective values obtained from the latter) although its bulk density was only 11% greater. The comparable relative air permeabilities (related to the slate composition using a pressure of 30 mm of water) suggest that the pore structures of the two materials were similar.

Another possible use for slate waste is in the production of autoclaved aerated concrete. (For the reader requiring background information, a recent CEB (Comité Euro-International du Béton) publication[18] discusses the manufacture and properties of autoclaved aerated concrete as well as detailed design and economic considerations involved in its use.) This type of concrete is a lightweight cellular material which is widely used in the construction industry, both in the form of blocks and as reinforced units. Its method of manufacture is fundamentally simple. For instance, in producing blocks, Portland cement (often with lime) is mixed with a fine siliceous material such as ground sand or pulverized fuel ash, also a small quantity of aluminium powder, and water to give a slurry; this is then poured into a large mould where the alkaline components react with the aluminium to form hydrogen gas bubbles and the slurry rises and sets to a cellular 'cake' which is then sliced, vertically and horizontally, into individual blocks which are autoclaved to give the finished material.

Watson, Eden and Farrant have used slate powder to make autoclaved aerated concrete on an experimental scale;[15] they selected the optimum 45:55 ratio of slate powder to Portland cement that had been established previously with materials from the same batches.[1] They varied the water/total solids ratio from 0·4 to 0·8 and the aluminium content from 0·0 to 0·6% (based on the weight of cement). The same autoclave cycle was used as previously (see Fig. 5.1) but, after removal from the autoclave, the specimens were air dried at 18–21 °C and 42–52% relative humidity for 4 days (to constant weight) before testing. Figures 5.6 and 5.7 show the dry densities and compressive strengths, respectively, of these materials plotted against aluminium content for the various water/total solids ratios used. The figures show that there was no apparent advantage in exceeding an aluminium content of 0·3% since any further increase hardly affected these properties.

The similar appearance of the two figures suggested plotting compressive strength against density, and this has been done in Fig. 5.8, which includes additional data obtained in subsequent work. Regression analysis gives very high correlation coefficients both for a parabolic relationship and for a

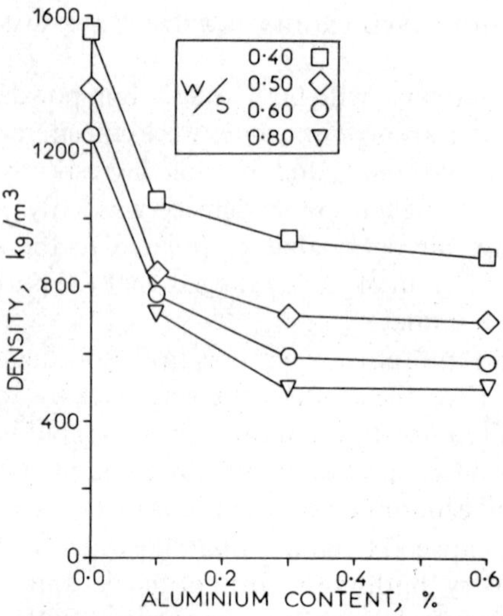

Fig. 5.6. The effect of water/solids ratio and aluminium content on the dry density of autoclaved slate/cement compositions (from ref. 15).

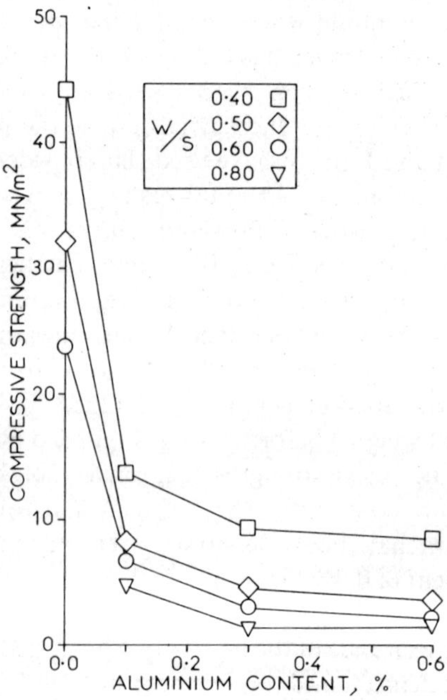

Fig. 5.7. The effect of water/solids ratio and aluminium content on the dry compressive strength of autoclaved slate/cement compositions (from ref. 15).

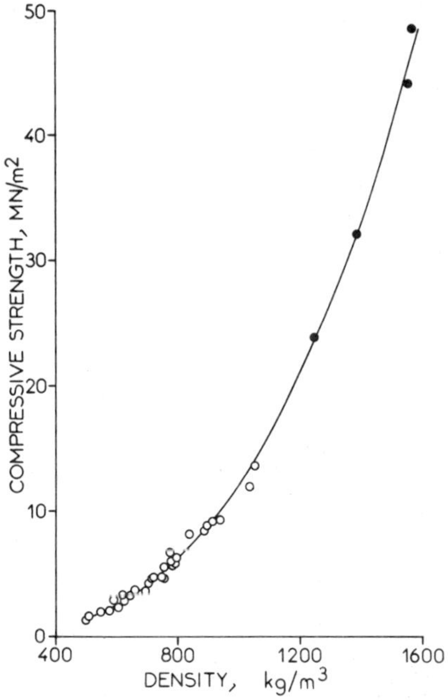

Fig. 5.8. The relationship between dry compressive strength and dry density of aerated and non-aerated autoclaved slate/cement compositions (aerated materials represented by open circles and non-aerated materials by solid circles).

power relationship. It seemed rather surprising that the non-aerated materials (i.e. with no aluminium powder added to the slurry), with their extremely fine porous structures, appear to fall along the same line as the aerated materials into which were incorporated very much coarser cellular structures. In fact, the extrapolations of the power regression lines for the non-aerated materials alone and for the aerated materials alone both fall very close to the power regression line for all the materials taken together over the whole density range that is shown in the figure. The actual expressions were found to be

(i) for the non-aerated materials alone:

$$f_c = (1{\cdot}19 \times 10^{-8})D^{3{\cdot}00} \qquad r = 0{\cdot}996$$

(ii) for the aerated materials alone:

$$f_c = (1{\cdot}47 \times 10^{-8})D^{2{\cdot}97} \qquad r = 0{\cdot}990$$

(iii) and for all the materials together:

$$f_c = (1{\cdot}26 \times 10^{-8})D^{2{\cdot}99} \qquad r = 0{\cdot}996$$

where f_c is the compressive strength (MN/m^2), D is the dry density (kg/m^3) and r is the correlation coefficient.

The power coefficient of 3 seems to be in reasonably good agreement with that found in commercial practice. For instance, Grimer and Brewer[19]

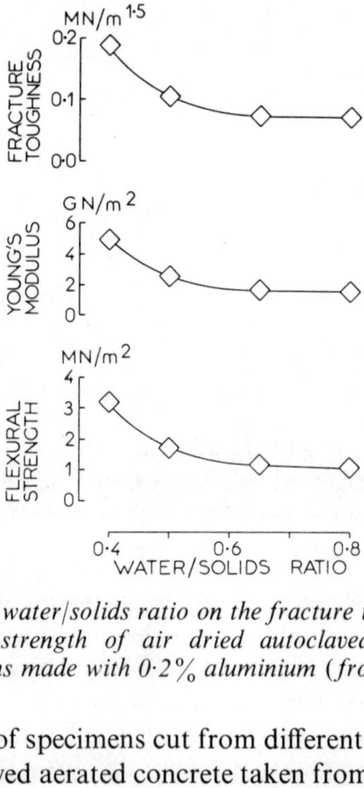

Fig. 5.9. The effect of water/solids ratio on the fracture toughness (K_c), Young's modulus and flexural strength of air dried autoclaved aerated slate/cement compositions made with 0·2% aluminium (from ref. 15).

measured properties of specimens cut from different positions in cakes of conventional autoclaved aerated concrete taken from the production lines of four different manufacturers. They found a general increase in specific gravity with depth in the cake (and hence in the mould) and this was accompanied by an increase in compressive strength. They analysed their results, using a power relationship like that shown above, and found that the data from three of the manufacturers' materials gave power coefficients between 2 and 3; for the fourth, they found a coefficient greater than 4 but in this case suspected that some other factor as well as specific gravity was influencing the strength.

Watson, Eden and Farrant measured the Young's modulus and flexural

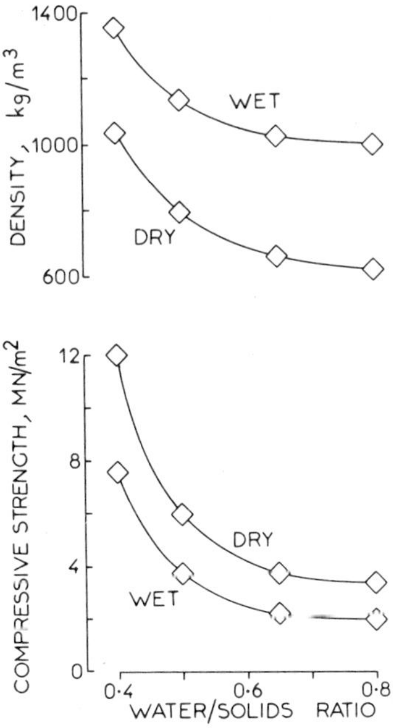

Fig. 5.10. The effect of water/solids ratio on the density and compressive strength of autoclaved aerated slate/cement compositions made with 0·2% aluminium (from ref. 15).

strength of some of the materials which they prepared and, as with the compressive strength, they found similar values to those quoted in the literature for commercial materials. (However, it should be noted that possible effects due to differences in test procedures were not taken into account; furthermore, the specimens which they prepared were relatively small (20 × 20 × 250 mm beams) because of the constraints imposed by the capacity of their autoclave and by the multifactorial experimental designs that they used.) They also estimated the fracture toughness (K_c) of some of their materials. To illustrate the range of values obtained, Figs. 5.9 and 5.10 show the effect of water/solids ratio (with 0·2% aluminium content) on fracture toughness, Young's modulus, flexural strength, density and compressive strength.

Toughness is an important property, particularly since some autoclaved aerated concretes are prone to damage during transport as well as due to handling on site; however, very little systematic work has been done on this aspect of their behaviour. Watson, Eden and Farrant estimated toughness using a measure of the critical stress intensity factor, K_c. However, it has been suggested that this approach should be regarded with caution when used with laboratory-scale specimens of hardened cement paste; linear elastic fracture mechanics, on which the method depends, may not be readily applicable to this type of material.[20] The same may also be true of these autoclaved compositions, which are broadly related to hardened cement paste, and Watson[21] has recently suggested that a more direct

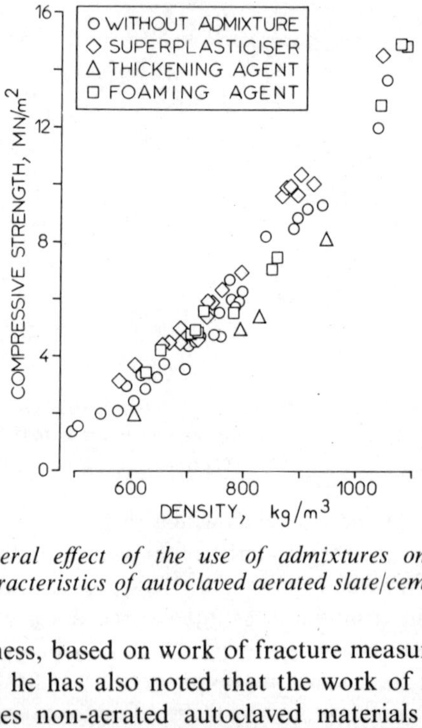

Fig. 5.11. The general effect of the use of admixtures on the compressive strength/density characteristics of autoclaved aerated slate/cement compositions.

estimate of toughness, based on work of fracture measurements, may be more appropriate; he has also noted that the work of fracture method clearly distinguishes non-aerated autoclaved materials from hardened cement paste (and there are considerable microstructural differences between the two) whereas a comparison based on linear elastic fracture mechanics does not.[22] Further discussion of this particular aspect is beyond the scope of the present book, although the toughness of slate-based autoclaved aerated concretes will be further considered later on.

Watson, Eden and Farrant also found indications that the addition of a commercial superplasticizer to their slurries had the effect of shifting the compressive strength/density curve, for the slate-based materials, to provide a greater compressive strength at a particular density; the addition of a thickening agent had the opposite effect. Furthermore, the cellular structure could also be produced by mixing the slate/cement blend with a stable water foam made with the aid of a foaming agent. Figure 5.11 shows the overall results that they obtained.

Details of this work are published elsewhere[16] but it is worth briefly examining some of the results because they have an important bearing on specimen preparation. Again, the optimum 45:55 ratio of slate powder to cement was used; and mixes were made which contained 3·5% (based on the weight of cement) of a commercial surfactant lignosulphonate superplasticizer and others with 0·03% of a commercial polyelectrolyte thickening agent, as well as plain mixes with no admixture. Aluminium contents of 0·1 and 0·3% were used (based on the weight of cement) together with water/total solids ratios of 0·5 and 0·7; and the autoclave cycle was the same as that shown in Fig. 5.1.

The relationship between dry compressive strength and dry density is shown in Fig. 5.12. The overall effects of aluminium content and water/solids ratio are as expected from Figs. 5.6 and 5.7, but superimposed on these are the shifts associated with the admixtures. The addition of the

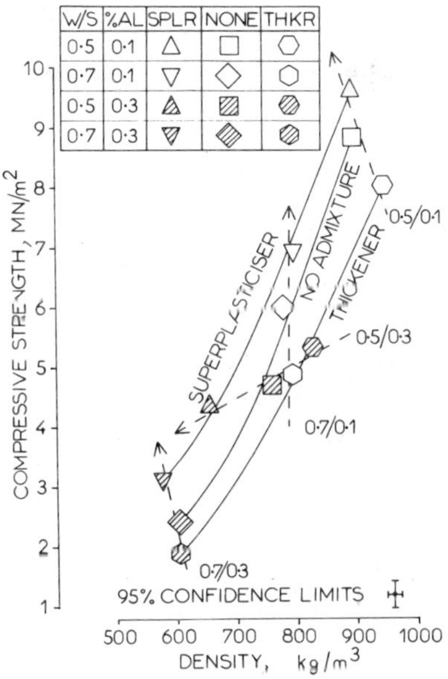

Fig. 5.12. Compressive strength/density relationships for autoclaved aerated slate/cement compositions made from superplasticized, plain and thickened slurries (from ref. 16).

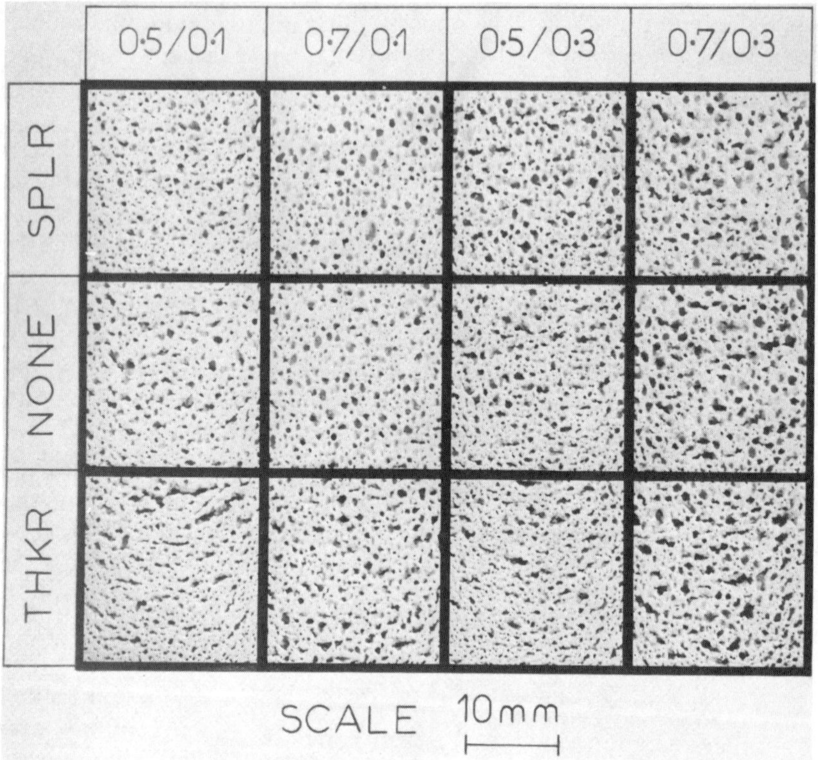

Fig. 5.13. Polished cross-sections through the materials indicated in Fig. 5.13, with the vertical direction as cast (from ref. 16).

superplasticizer has a beneficial effect on the strength/density relationship whereas the thickener has an adverse effect. The arrows indicate the progression of the strength/density characteristic, for a given mix, from thickened to plain to superplasticized slurries; the 0·5/0·3 % materials became less dense, and weaker, along this series whereas the others became stronger with relatively little change in density.

Figure 5.13 shows the cellular structure of the materials. The 0·5/0·3 % mixes gave an increasingly expanded structure, in progressing from the thickened to the plain to the superplasticized mixes, in accordance with the density reduction, and there was a correspondingly greater uniformity in the formation of the cellular structure. (It seems that the potentially high expansion, with 0·3 % aluminium, tends to be restrained by the 'dryness' of the mix, and that this effect is exaggerated by the thickener and alleviated by the superplasticizer.) The other materials also show a corresponding improvement in uniformity, which accompanies their increase in strength (and this is particularly evident in the case of the 0·5/0·1 % materials where the thickened mix gives the appearance of having been 'torn' during the expansion process). In general, these other mixes seem to be fluid enough for the potential expansion to be virtually unrestrained; the effect of the superplasticizer aids the formation of a uniform cellular structure, or

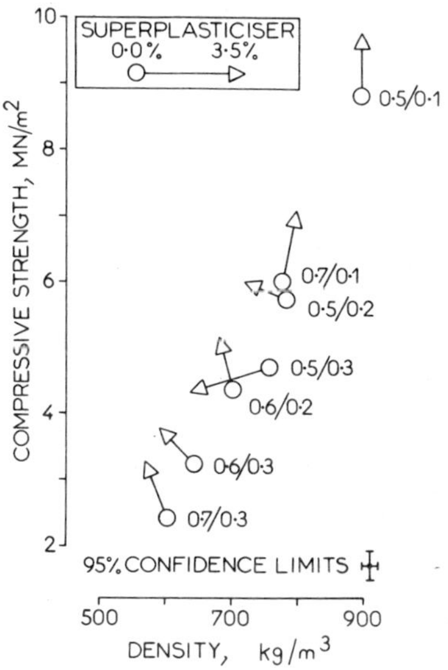

Fig. 5.14. Strength/density shifts associated with the addition of 3·5 % superplasticizer (from ref. 16).

perhaps allows more or less full expansion to take place before the onset of setting which would otherwise cause tearing of the expanding structure—it may also improve the interdispersion of the slate and cement grains, thus aiding the formation of reaction products during autoclaving, and perhaps the aluminium powder too.

The effect of superplasticizer addition has proved to be a useful diagnostic aid in experimental mix design. Figure 5.14 shows the shift 'vectors' associated with the addition of 3·5 % superplasticizer to the mix. Three intermediate mixes (0·5/0·2 %, 0·6/0·2 % and 0·6/0·3 %) were prepared to provide additional information. To aid interpretation, Fig. 5.15 shows the shift vectors arranged according to water/solids ratio and aluminium content. It is clear that there is a progressive transition from the density-reducing to the strength-increasing role of the superplasticizer as the aluminium content decreases and the water/solids ratio increases. A more or less vertical vector suggests that the corresponding plain mix is of the right composition to achieve its full potential expansion. Mixes prepared in subsequent work by Watson and his co-workers have been selected on the basis of the vertical vector criterion in order to minimize

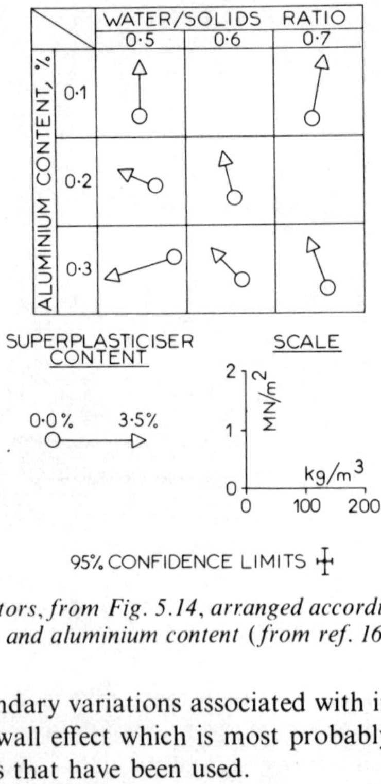

Fig. 5.15. Shift vectors, from Fig. 5.14, arranged according to water/solids ratio and aluminium content (from ref. 16).

errors due to secondary variations associated with incomplete expansion and with the side-wall effect which is most probably significant with the small-sized moulds that have been used.

Watson, Eden and Farrant also described work using different superplasticizer contents[16] and this confirmed that the interaction between the superplasticizer and the aluminium and water contents is complex. For instance, Fig. 5.16 shows the effect of varying the superplasticizer and aluminium contents simultaneously at a constant water/solids ratio of 0·5. The plotted lines, at each aluminium content, represent the progressive strength/density shift with increasing superplasticizer; and these were obtained by combining separate plots of strength and density against

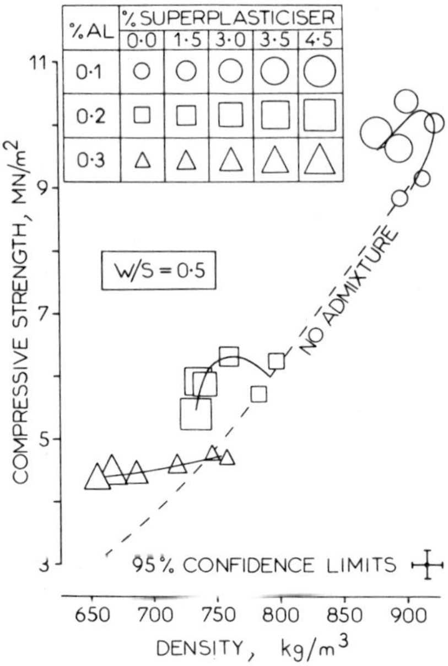

Fig. 5.16. Strength/density shifts associated with varying superplasticizer and aluminium contents at a water/solids ratio of 0·5 (from ref. 16).

superplasticizer content. The figure suggests that there may be a transition from the strength-increasing role to the density-reducing role at higher superplasticizer contents with a relatively small proportion of aluminium. (It should be noted that the results as a whole suggest that the strength-increasing role of the superplasticizer still tends to persist even when the density-reducing role appears to predominate; at a given density, a material made with superplasticizer is generally stronger than one made without, even when the strength/density characteristic has been shifted by a reduction in density.)

As noted above, Watson, Eden and Farrant also formed the cellular structures by mixing the slate/cement blend into a preformed stable water foam (using a sulphated lauryl alcohol foaming agent at a concentration of 1·2% based on total solids); water/solids ratios of between 0·4 and 0·7 were used and the corresponding cellular structures are shown in Fig. 5.17. The two higher water/solids ratios gave slurries that poured readily but it was difficult to produce castings free of large voids with the others. After setting, further preparation and conditioning followed the same routines adopted for the materials made with aluminium powder

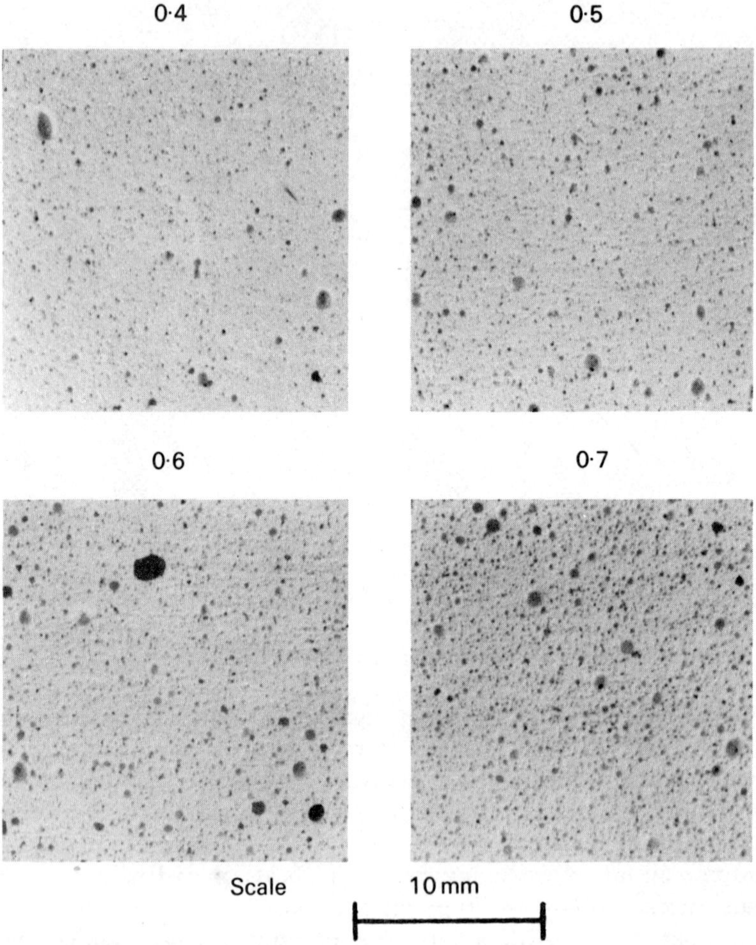

Fig. 5.17. Polished cross-sections through autoclaved aerated slate/cement compositions made from foaming agent at water/solids ratios of 0·4–0·7 (from ref. 16).

As the water/solids ratio was increased, the strength and densities were progressively reduced and, as the figure shows, the cellular structures tended to become coarser although in all cases they were very much finer than those produced with aluminium powder. Figure 5.18 shows that the strength/density characteristics were very similar to the aluminium-expanded materials and at lower densities their strengths tended towards

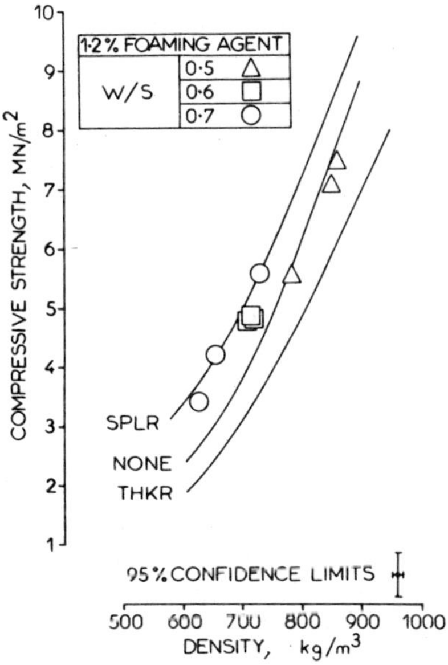

Fig. 5.18. Compressive strength/density characteristics of autoclaved aerated slate/cement compositions made with foaming agent, compared with the strength/density relationships for the aluminium expanded materials (from ref. 16).

those using the superplasticizer; however, the strength/density characteristics of different specimens from the same mix often showed considerable scatter that was presumably associated with the method of preparation.

There has been very little work done on the effect of slate particle size distribution on the properties of slate-based autoclaved aerated concrete. However, an investigation, by Hickson,[17] examined the effect of mixing different proportions of fine and coarse slate powders (of which the lower ends of their respective particle size distributions are indicated in Table 5.7). The mixes he used consisted of 70% of the slate powder blend, 15% ordinary Portland cement, 15% ground quicklime and 0·05% aluminium powder (based on total solids). Sufficient water was added to give a constant viscosity, which was determined by means of a flow test not unlike the slump test used for concrete mixes; and 3–4 hours after casting, the specimens were trimmed and then autoclaved at about 1·03–1·07 MN/m^2 for nearly 8 hours. Table 5.8 shows the characteristics of four mixes containing varying proportions of the fine and coarse slate powders. The table indicates that, as the proportion of the coarser powder was increased, less water was needed to give the required viscosity; furthermore, there was a concomitant

Table 5.7. Particle size distribution, below 0·150 mm, of slate powders blended for use in experimental autoclaved aerated concretes (from ref. 17).

Sieve size (mm)	*% passing*	
	Fine slate powder	*Coarse slate powder*
0·150	99·8	34·0
0·090	92·5	20·8
0·063	44·0	11·2
0·053	27·5	7·0

increase in the hardening rate and a decreasing tendency for the expanded cake to sink. The increasing proportion of coarser powder also tended to result in a higher dry density of the autoclaved product but, at the same time, the saturated compressive strength fell. The reasons for the decreasing strength are not clear, especially in view of the lower water/solids ratios and the higher densities, but they may result from the lower specific surface area of the slate powder and the consequent reduction in the quantity of reaction products formed.

Eden[23] has recently examined some engineering properties of slate-based

Table 5.8. Characteristics of experimental mixes for autoclaved aerated concrete and of the finished materials, using blends of fine and coarse slate powder (from ref. 17).

Mix characteristics			*Characteristics of autoclaved materials*	
Slate blend	*Water/solids ratio*	*Hardening characteristics*	*Compressive strength (wet) (MN/m^2)*	*Dry density (kg/m^3)*
100% fine	0·81	Excessive sinking. Slow hardening rate.	2·34	597
75% fine, 25% coarse	0·67	Slight sinking.	2·25	574
50% fine, 50% coarse	0·57	No sinking. Satisfactory hardening rate.	1·93	649
25% fine, 75% coarse	0·50	Fairly rapid hardening rate.	1·90	662

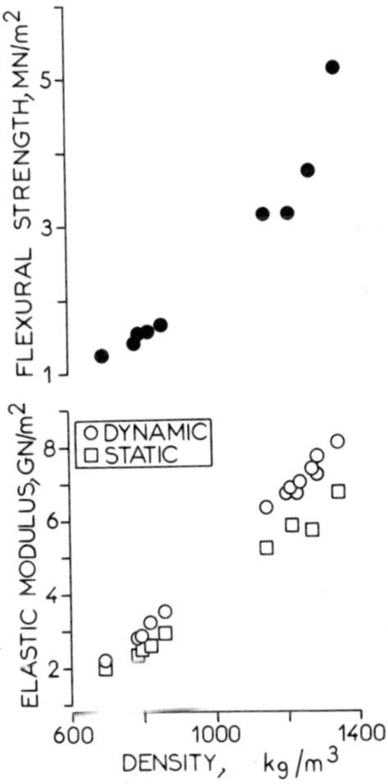

Fig. 5.19. Flexural strength and dynamic and static elastic moduli of some autoclaved slate/cement compositions plotted against density (from ref. 23).

autoclaved aerated concrete in detail and has begun to relate these to their microstructural characteristics. Initially he used the 45:55 slate/cement ratio, with similar materials to those employed in the previous investigations,[1,15,16] and varied the water/total solids ratio and the aluminium content in accordance with the vertical vector criterion discussed above. The materials were normally autoclaved, 2 days after casting, using a cycle very similar to that shown in Fig. 5.1.

His flexural strength and modulus as well as dynamic elastic modulus results are plotted against density in Fig. 5.19; the specimens were tested after they had reached moisture equilibrium at 45–50 % relative humidity and 21 °C. The flexural load/deflection plots tended to be linear over most of their range up to failure. Furthermore, the relationship between dynamic modulus and flexural modulus was found to be approximately linear, and the dynamic values are generally about 11–28 % higher than the corresponding static values.

Figure 5.20 shows that the relationship between fracture surface energy, determined by the work of fracture method,[21] and density is complex. (It

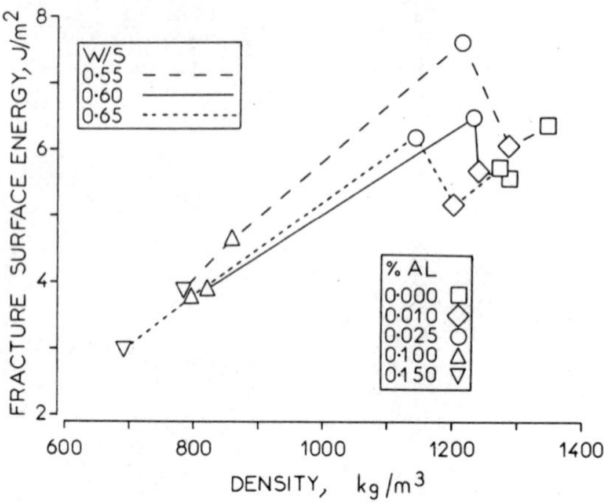

Fig. 5.20. Apparent fracture surface energy of some autoclaved slate/cement compositions plotted against density (from ref. 23).

should be noted that the energy values shown are based on the gross cross-sectional area of the specimen; they do not take into account the voids in the cellular structure.) Eden has associated the peak in fracture surface energy, at relatively low aeration (i.e. 0·025% aluminium), with the increased tortuosity of the propagating crack around the voids; as the density decreases below the peak, the effect of the reduced net cross-sectional area of the matrix material predominates and the fracture surface energy decreases accordingly.

Figure 5.21 shows the drying shrinkage associated with various combinations of water/solids ratio and aluminium content; the values were determined by soaking the specimens in water for 7 days and then

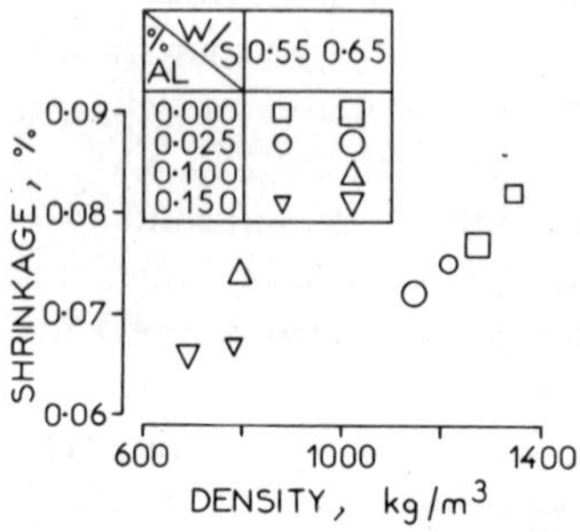

Fig. 5.21. Drying shrinkage of some autoclaved slate/cement compositions plotted against density (from ref. 23).

measuring the shrinkage as they were dried to equilibrium over saturated potassium carbonate solution (giving a relative humidity of about 43 %) in accordance with a recently recommended procedure.[24] The figure suggests that there is a slight reduction in shrinkage with increasing aeration; and this in turn suggests that there may be a change in the microstructure of the hydrothermal reaction products which is associated with increasing aeration.

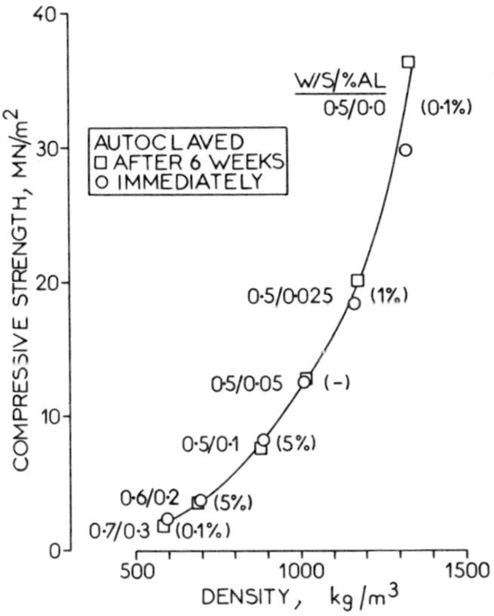

Fig. 5.22. Compressive strength/density relationships for some slate/cement compositions autoclaved immediately after demoulding and after 6 week's wet precuring (from ref 23). The level of significance of the difference in compressive strength mean values, for each mix, is shown in brackets.

Other evidence for microstructural differences associated with aeration is given in Fig. 5.22. This shows the compressive strength/density relationship for materials made from two sets of specimens that were identical apart from the fact that one was autoclaved immediately after demoulding and the other was autoclaved after 6 weeks' water curing. (The strength and density values, which are the means of six determinations, were obtained after drying at 40–50 % relative humidity at about 20 °C.) The figure includes the results of statistical tests on the pairs of compressive strength results for each mix; thus, in the case of the 0·7/0·3 % mix, 6 weeks' water curing gave a very slight but highly significant reduction in strength whereas, for the 0·5/0·0 % and 0·5/0·025 % mixes, there is a corresponding significant increase in strength. Since the differently cured sets originated from the same mixes, and therefore had virtually identical cellular structures for corresponding mix proportions, these effects most probably stem from microstructural differences.

In a preliminary investigation, Eden has shown that aeration modifies the volume of smaller pores in the microstructure of the matrix materials; the reason for this is not clear, although it may possibly be connected with the

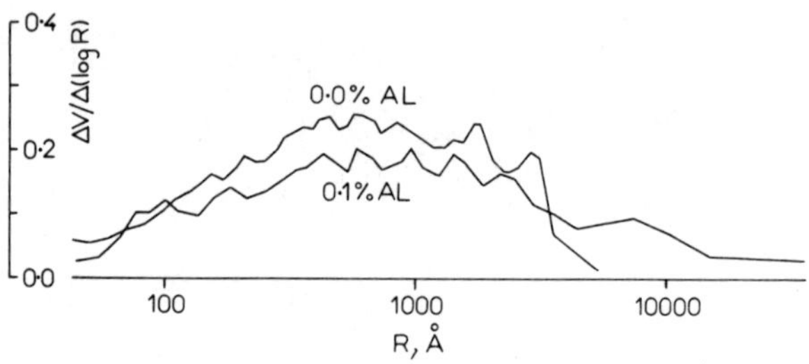

Fig. 5.23. The effect of aeration on the pore size distribution of an autoclaved slate/cement composition (from ref. 23).

greater accessibility to the interior of the 'green' material by the steam which is needed for the hydrothermal reactions to take place. Whatever the explanation, Fig. 5.23 shows the pore size distributions, determined by mercury pressure porosimetry, for materials prepared with a water/solids ratio of 0·55 and aluminium powder contents of 0·0 and 0·1 %; in this case, aeration is accompanied by a clear reduction in the volume of pores over a considerable range of relatively small equivalent pore radii, and the creation of others that are larger. Eden has also shown that, for both aerated and non-aerated materials, an increase in water/total solids ratio leads to an increase in porosity but that the pore size distribution in this range is not significantly affected.

Of course the autoclaving conditions have a marked effect on the pore size distribution. Figure 5.24 shows the effect of two different autoclaving cycles (119 °C for 15·5 hours and 170 °C for 8 hours) and of varying the slate content (0 %, 45 % and 55 %). Preliminary X-ray diffraction studies have indicated that, at 170 °C, the quartz present in the slate reacts with the lime from the cement but that the other slate minerals (chlorite, muscovite/illite) seem to be relatively unaffected; the reaction leads to the formation of a relatively crystalline calcium hydrosilicate, 11 Å tobermorite. However, the role of 11 Å tobermorite in determining the engineering properties is

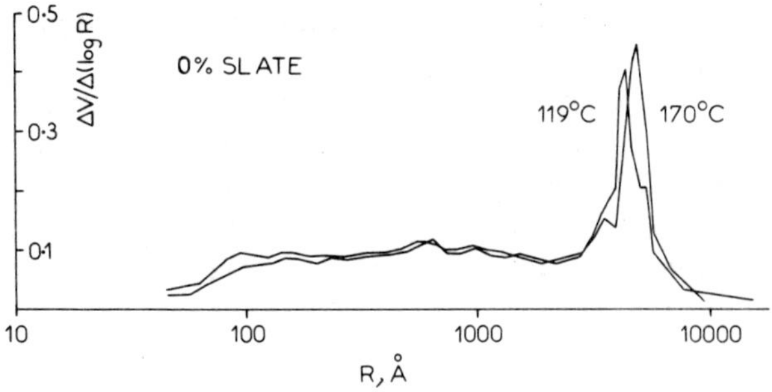

(a)

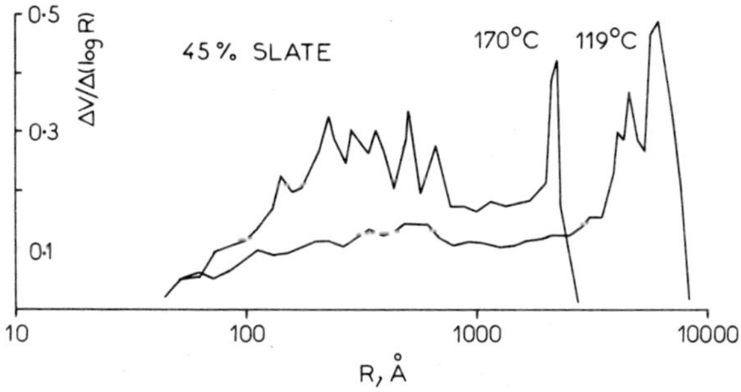

(b)

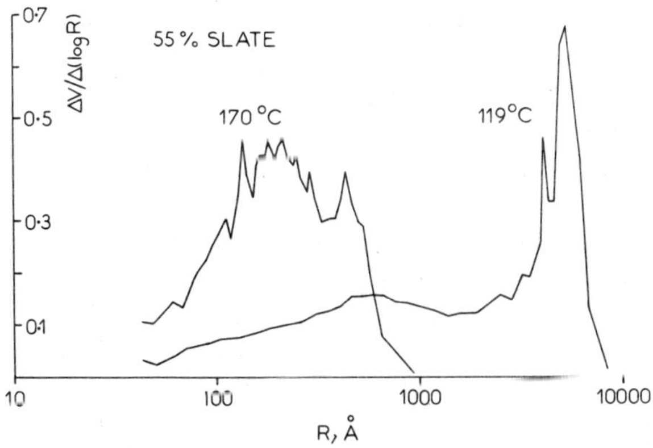

(c)

Fig. 5.24. The effect of autoclaving conditions (119°C for 15·5 hours and 170°C for 8 hours) and slate content (0, 45 and 55%) on the pore size distribution of some non-aerated autoclaved slate/cement compositions using a water/solids ratio of 0·5 (from ref. 23).

complex, and may differ for aerated and non-aerated materials respectively.[8]

As noted in the introduction to this chapter, the presence of suitably reactive Al^{3+} may increase the 'crystallinity' (i.e. the relative proportion of 11 Å tobermorite) of these autoclaved calcium hydrosilicate compositions. However, the apparent chemical inertia of the other minerals in the slate compared with the quartz seems to suggest, at this stage of the

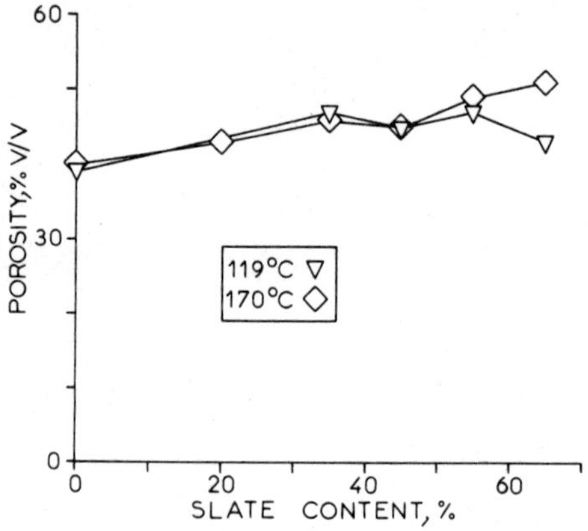

Fig. 5.25. The effect of autoclaving conditions and slate content on the helium porosities of some non-aerated autoclaved slate/cement compositions with a water/solids ratio of 0·50 (from ref. 23).

investigation, that the combined aluminium in the slate minerals is having little effect on the hydrothermal reactions;[25] and this could perhaps help to explain the surprisingly high drying shrinkage that has been found for slate/lime bricks.

Figure 5.24 shows that as the slate content increases, and the amount of quartz available for reaction increases, there is a marked change in pore structure on autoclaving at 170 °C; the formation of 11 Å tobermorite is accompanied by a large general reduction in the size of the pores although, as Fig. 5.25 indicates, the change in total porosity of non-aerated specimens (determined by helium pycnometry) is relatively small.

In the light of his findings, Eden has briefly examined the effect of slate content on density, compressive strength and dynamic elastic modulus, for aerated and non-aerated materials, using the two autoclave cycles; his results are shown in Fig. 5.26. After autoclaving for 15·5 hours at 119 °C, the slate appears to act as an inert filler; the properties of the non-aerated

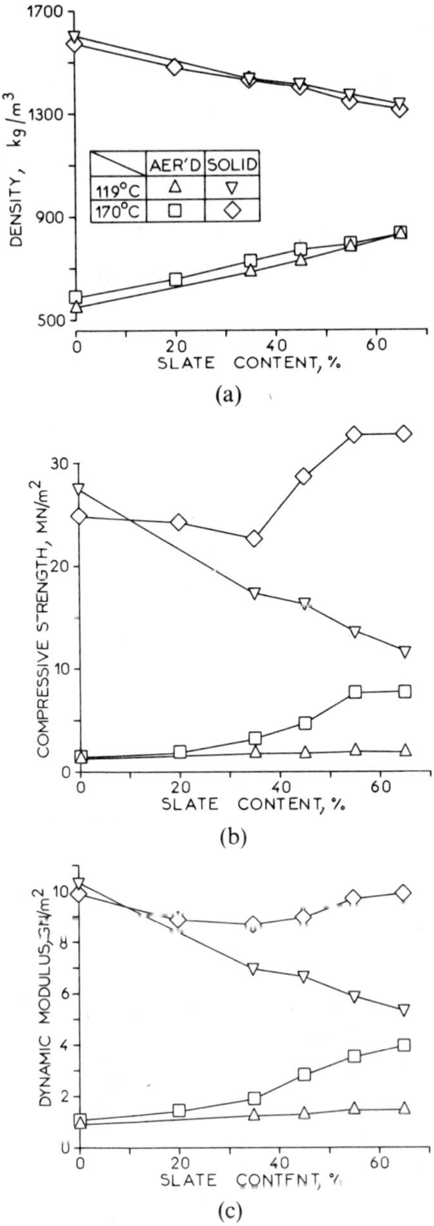

Fig. 5.26. The effect of autoclaving conditions and slate content on the density, compressive strength and dynamic elastic modulus values of some autoclaved slate/cement compositions (from ref. 23). (For the non-aerated materials, water/solids ratio = 0·50, and for the aerated materials, water/solids ratio = 0·65 with an aluminium content of 0·15%.)

mixes are progressively degraded as the slate content is increased. The corresponding slight improvements for the aerated materials appear to be associated with the increasing density; this density increase seems to be due to the stiffening action of the slate powder on the fresh slurry—as the slate content is increased, the expansion process is progressively restrained. After autoclaving for 8 hours at 170 °C, strength and modulus are generally markedly improved at higher slate contents, for both aerated and non-aerated materials, although the dynamic modulus of non-aerated 100% Portland cement seems to be relatively high.

At present, the results of Eden's work seem to suggest that the formation of 11 Å tobermorite tends to subdivide the porous structure, with a concomitant increase in compressive strength and dynamic modulus. However, much work remains to be done and it is hoped that full conclusions will be published in due course.

REFERENCES

1. K. L. Watson, R. D. Grant and R. M. Jones, 'A potential use for slate waste?', *Precast Concrete*, **7**(9) (Sep. 1976), 449–52.
2. R. H. Boyle (Director, Penrhyn Quarries Ltd), private communication.
3. C. A. Menzel, 'Strength and volume change of steam-cured Portland cement mortar and concrete', *Journal of the American Concrete Institute, Proceedings*, **31**(2) (Nov.–Dec. 1934), 125–48. (Reprinted in *Menzel Symposium on High Pressure Steam Curing*, Publication SP-32, American Concrete Institute, Detroit (1972), 221–44.)
4. *The Chemistry of Cements* (Editor: H. F. W. Taylor), Academic Press, London (1964), Volume 1, Chapter 11.
5. F. M. Lea, *The Chemistry of Cement and Concrete*, 3rd Edition, Edward Arnold (Publishers) Ltd, London (1970).
6. H. F. W. Taylor, 'A review of autoclaved calcium silicates', *Symposium on Autoclaved Calcium Silicate Building Products, London*, May 1965, Society of Chemical Industry, London (1967), 195–205.
7. G. Verbeck and L. E. Copeland, 'Some physical and chemical aspects of high pressure steam curing', *Menzel Symposium on High Pressure Steam Curing*, Publication SP-32, American Concrete Institute, Detroit (1972), 1–13.
8. J. M. Crennan, S. A. S. El-Hemaly and H. F. W. Taylor, 'Autoclaved lime–quartz materials: 1. Some factors influencing strength', *Cement and Concrete Research*, **7**(5) (1977), 493–502.
9. A. Noorlander, 'Reducing the shrinkage of calcium silicate bricks', *Symposium on Autoclaved Calcium Silicate Building Products, London*, May 1965, Society of Chemical Industry, London (1967), 152–4.
10. S. M. Rozenblit, *Stroitel. Materialy*, No. 2 (1938), 4–19. (*Chem. Abstr.*, **34** (1940), 7558^1.)

11. US 2,741,798 (1956). (*Chem. Abstr.*, **50** (1956), 13400c.)
12. E. M. Lawson and P. J. Nixon, *A Survey of the Locations, Disposal and Prospective Uses of the Major Industrial By-products and Waste Materials in Scotland*, Building Research Establishment Current Paper CP 50/78, Department of the Environment (Jun. 1978).
13. *Properties and New Uses of Pennsylvania Slate*, Mineral Industries Experiment Station Bulletin No. 47, Pennsylvania State College (1947).
14. *Experimental Investigation on Slate Waste from Aberfoyle Quarry for Use in Slate–Lime Bricks*, Building Research Station Note No. A 41, Department of Scientific and Industrial Research (Nov. 1954).
15. K. L. Watson, N. B. Eden and J. R. Farrant, 'Autoclaved aerated materials made from slate powder and Portland cement', *Precast Concrete*, **8**(2) (Feb. 1977), 81–5.
16. K. L. Watson, N. B. Eden and J. R. Farrant, 'The effect of admixtures on the relationship between compressive strength and density of autoclaved aerated concrete made from slate powder and Portland cement', *Silicates Industriels*, **43**(3) (Mar. 1978), 57–63. (Paper presented to the *International Conference on Cement and Concrete Admixtures and Improving Additives, Mons*, Nov. 1977.)
17. G. W. Hickson (Technical Adviser, Aerated Concrete Ltd), private communication.
18. 'Autoclaved aerated concrete', *CEB Manual of Design and Technology*, Comité Euro-International du Béton, The Construction Press Ltd, Lancaster (1978), 90 pp.
19. F. J. Grimer and R. S. Brewer, 'The within-cake variation of autoclaved aerated concrete', *Symposium on Autoclaved Calcium Silicate Building Products, London*, May 1965, Society of Chemical Industry, London (1967), 163–70.
20. D. D. Higgins and J. E. Bailey, 'Fracture measurements on cement paste', *Journal of Materials Science*, **11**(11) (Nov. 1976), 1995–2003.
21. K. L. Watson, 'The estimation of fracture surface energy as a measure of the "toughness" of hardened cement paste', *Cement and Concrete Research*, **8**(5) (1978), 651–6.
22. K. L. Watson, 'Reply to a discussion of "The estimation of fracture surface energy as a measure of the 'toughness' of hardened cement paste" by Y. W. Mai', *Cement and Concrete Research*, **9**(4) (1979), 541–4.
23. N. B. Eden (Department of Civil Engineering, Portsmouth Polytechnic), unpublished work.
24. 'Tentative recommendations: aerated concrete test methods', *Materials and Structures, Research and Testing*, **8**(45) (May–Jun. 1975), 211–33.
25. Also see discussion in ref. 9.

6

Miscellaneous Applications for Slate Waste

6.1. INTRODUCTION

Previous chapters have been concerned with slate products that are already commercially established (such as granules, powder, expanded slate and, to some extent, crushed slate for concrete and road construction) or applications that appear to have obvious commercial potential, particularly the manufacture of autoclaved products. This chapter is primarily concerned with miscellaneous applications that have been investigated on a limited scale but which do not seem to be of such widespread or immediate commercial significance. These include the manufacture of mineral wool, fired bricks, alumina and aluminium, glass products and other possibilities; of course, some of these have been exploited commercially although not necessarily with any large degree of success.

6.2. MINERAL WOOL

Mineral wool consists of fine inorganic fibres and is commercially available in a variety of forms.[1] It can be obtained loose and in the form of blankets (which may be covered on one or both sides with paper, cloth or aluminium foil) as well as mattresses with cloth or wire netting; it is also marketed as semi-rigid felted slabs, as resin or bitumen bonded boards and it can be incorporated into insulating cements. Obviously, one of its more significant applications is for thermal insulation in buildings; for these applications it can be used in prefabricated form, and there are techniques which enable loose wool to be blown into cavities. It is also used for both hot and cold insulation of vehicles as well as plant and equipment. Mineral wools also have useful sound-deadening properties; and they have been used as a packing material and some types have been used as a filtration medium.[2]

A variety of raw materials, including certain slags and a number of natural minerals, can be used to manufacture mineral wool. In principle, the process generally involves pouring a stream of molten material into a jet of steam or air;[1,2] this results in the formation of fine droplets which, as they are rapidly propelled by the jet stream, leave fine thread-like trails of material in their wake. The mass of thread-like fibres, which constitutes the wool, may contain a proportion of 'shot' particles which can be removed; and some commercial products contain a certain amount of shot. Additives, such as waterproofers or disinfectants, can be applied at the jet. The raw materials used, the melt temperature and flow, and the factors affecting the jet of air or steam will influence the quality and dimensions of the fibre and the content of shot.

A report by the Building Research Station in the United Kingdom[1] noted that mixtures of rocks or slags, with melting points between 1350 and 1400 °C and viscosities of not more than 5 P at 1500 °C or 15 P at 1400 °C, can be used to make mineral wool. The report quoted the usual ranges of composition (after F. E. Lathe) as 35–45 % silica, 20–40 % lime (CaO), 10–20 % alumina, 2–15 % magnesia (MgO), 0–2 % alkalies and 1–3 % iron oxide (FeO); corresponding extreme ranges were given as 25–65 %, 10–45 %, 0–25 %, 0–20 %, 0–5 % and 0–20 % respectively, and up to 0·75 % sulphur.

It was noted that the silica is essential, the iron oxide and sulphur (and manganese dioxide) undesirable, and the other components desirable.

The mineralogy of slate suggests the possibility of using it as a raw material for the manufacture of mineral wool. This observation led to a laboratory-scale investigation which was undertaken as part of the Mineral Industries Experiment Station programme.[2] Four samples of commercial rock wool were analysed and their silica content was found to range between 36·9 and 44·0 %. For three of the samples, the range of alumina contents was 10·0–19·5 %, iron oxide (Fe_2O_3) 1·0–1·9 %, lime (CaO) 26·0–41·8 % and magnesia (MgO) 0·6–19·0 %; and, in comparison, the fourth sample had very low alumina and lime contents and a very high iron oxide content. There were also minor quantities of other materials, including sulphur in one case.

A laboratory-scale production method was developed with a steel mill slag that was actually used as a raw material for commercial production of mineral wool. Crushed slate was then screened to below 10 mesh, mixed with various quantities of limestone and calcined at about 1100 °C for 1–1½ hours in a gas-fired furnace to remove volatile material; the melt temperature was then increased to 1430 °C and the material poured into a stream of compressed air which formed the fibres and carried them on to a wire collecting screen. Limestone contents of 0, 22 and 44 % resulted in a progressively lower melt viscosity. Zero limestone gave an unpourable melt and 22 % produced thick fibres with strongly attached shot particles; 44 % gave fine, flexible, pale-coloured fibres, with lightly attached shot particles, but they were short and fragile. Some of the best results were achieved using a mixture of equal quantities of slate and steel mill slag.

Table 6.1. The composition of some slate-based mixes used for producing experimental mineral wool (from ref. 2).

	A	*B*	*C*	*D*	*E*
Slate	60	60	55	65	60
CaO	30	20	20	25	25
MgO	10	20	20	10	10
SiO_2			5		5

It was recognized that the different oxides in the melt needed to be properly balanced, and calcium and magnesium carbonates (in the form of limestone and dolomitic limestone) were added to the slate mixes. Work was also carried out with a eutectic mixture which gave a good quality wool.

Taking all the accumulated information into account, a final set of mixes was prepared and their compositions are shown in Table 6.1. Mixes A, B and C were calcined at 1100–1200 °C for 2 hours and then poured and blown after the melt temperature had been raised to 1350 °C; mix A gave the best results, the silica and higher magnesia contents of the others probably reducing the quality of their fibres. (It was noted that better textured fibres were produced from the melt poured first; poorer quality fibres were obtained as the melt cooled and became more viscous.) Mixes D and E were calcined and heated to about 1370 °C for pouring. Mix E was too viscous and gave coarse, dark-coloured fibres; however, mix D was one of the best, using the highest proportion of slate and producing long, fine, flexible fibres.

Shortly after the Mineral Industries Experiment Station report was published, there was a proposal to manufacture slate wool from the waste produced at the Dinorwic Quarries in North Wales[3] and there have been further investigations since then.[4,5] Phillips[5] found that X-ray studies of a particular slate wool indicated an amorphous structure. Figure 6.1 shows a photograph of this material, and Fig. 6.2 shows scanning electron micrographs of individual fibres at different magnifications. Thermal

Fig. 6.1. Slate wool (by courtesy of Dr G. O. Phillips).

analysis indicated a low moisture content and a range of thermal stability which extended up to 600 °C; furthermore, the material exhibited a small, continuous linear expansion up to 675 °C. It showed no significant deterioration after several days' exposure to water and to mild acids and alkalies. After these preliminary studies, it was suggested that further evaluation should include the investigation of properties related to the behaviour of slate wool as reinforcement in composite materials.

In conclusion, it should perhaps be noted that the Building Research Station report[1] made the point that slate waste has no particular merit for making mineral wool apart from the fact that the material has already been extracted and is available in roughly crushed form. On the basis of present knowledge, there does not seem to be evidence that it has outstanding characteristics which particularly distinguish it from more conventional raw materials used for making mineral wools.

6.3. FIRED BRICKS

The possibility of making fired bricks from slate waste has been investigated by the Rural Industries Bureau (now CoSIRA, the Council for Small Industries in Rural Areas) in the United Kingdom; a report, written in 1962, summarized the work that had been done up until that time.[6] This work was

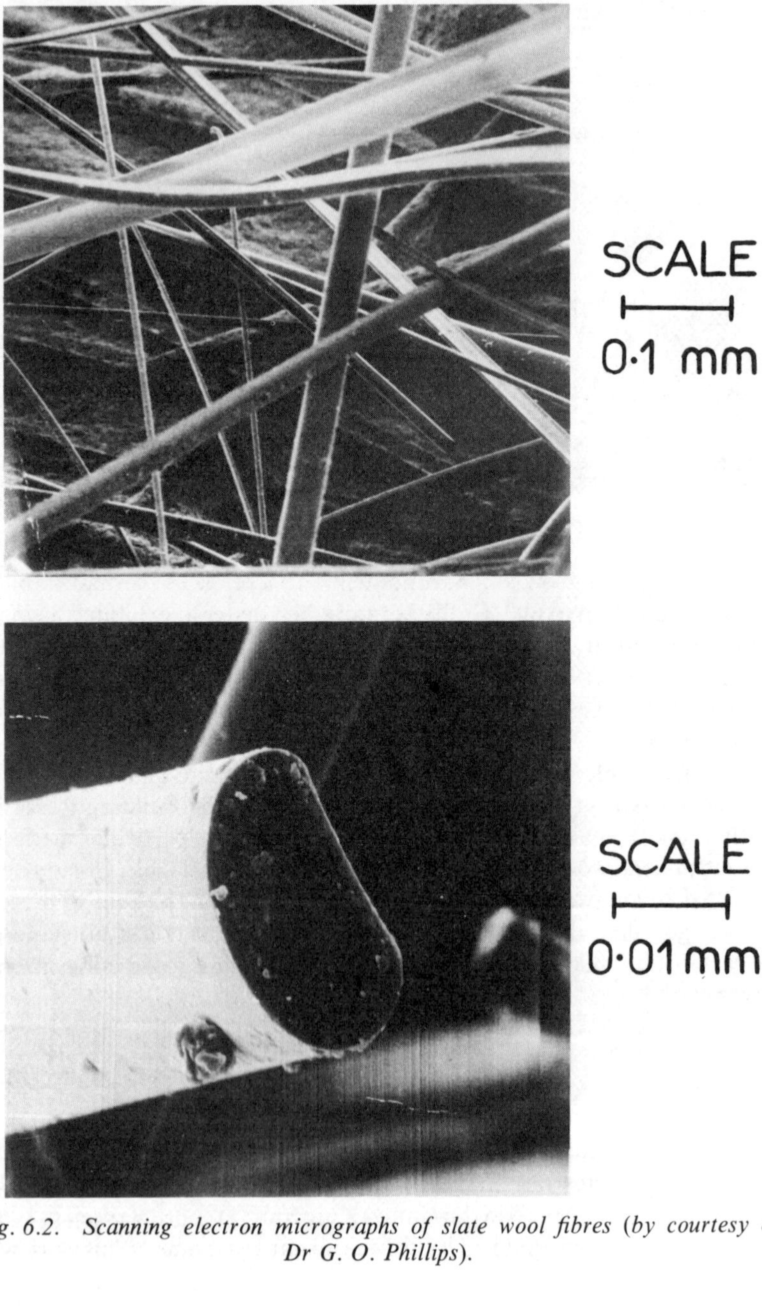

Fig. 6.2. Scanning electron micrographs of slate wool fibres (by courtesy of Dr G. O. Phillips).

prompted by the slate waste problem and by unemployment in the Nantlle district in North Wales. Initial experimental work began in 1946, using more or less orthodox brick making methods; crushed and screened slate waste was moulded and fired at 1100 °C and the results suggested that high grade bricks could be produced by proper processing methods.

The work was developed further, and slate samples were obtained from quarries throughout Wales and also from Scotland and Cornwall. Examination of the samples indicated differences but generally, after satisfactory gradings had been established, none of the materials from the different sources gave significant problems—and bricks made from the various wastes appeared to be very similar. The effect of different burning cycles was also investigated; the rate at which the temperature was increased and the final temperature were both varied. (N.B. Clearly, in making fired bricks from slate waste, the combination of raw material and processing conditions must be adjusted to avoid problems associated with 'bloating'; this is in contrast with the production of expanded slate aggregate, discussed in Chapter 3, where bloating is essential to provide a lightweight cellular structure.)

The Rural Industries Bureau studies showed that most of the slate wastes from Wales that were investigated could be used on the pilot scale to produce bricks with high crushing strength, very low porosity and high acid resistance, as shown by British Standard tests; however, it was noted that the standard brick making process would have to be modified slightly.

The report pointed out that slate waste has some advantages over clay (and also over marl or shale). Its water absorption is low; furthermore, it is not 'baked' by the sun and most of the problems encountered in the operation of a clay pit would be avoided. From the grinding and screening point of view, its low water absorption would be an advantage in maintaining output under different weather conditions, and its use would reduce the problem of 'blinding' the screens which is encountered with clay. Its moisture retention characteristics would result in both easier control of the proper consistency for pressing and in the need for only a slight amount of drying. Furthermore, the drying shrinkage was found to be small; the shrinkage was reflected in a low degree of movement during burning, and products could be made to much finer limits.

However, the report pointed out that there were disadvantages with slate. For instance, a large primary crusher might be needed and there might be problems with wear on fine grinding plant; furthermore, facilities for dust extraction would be required. An efficient method was needed for preparing the damped dust before pressing, and the slate mixtures did not have the 'plasticity' of clay. Finally, it was noted that the 'green' bricks needed careful handling, but that there were no apparent disadvantages in either the drying or burning stages of the process.

The report also included some properties of bricks made from slate. Crushing strengths of 105·5 and 95·7 MN/m^2 (with corresponding densities of 2140 and 2050 kg/m^3 respectively) were quoted for some full-sized bricks that had been fired in an electric kiln at 1120 °C for 55·5 hours. Some were tested for their resistance to acids by a British Standard method, BS 784:1953 (which gave values of between 98·7 and 99·3 %). The report also referred to tests made on other slate bricks that had been exposed to the corrosive conditions in a sewer for 2 years; half of the samples had been in the atmosphere at the top of the sewer and half of them had been immersed. After exposure, the bricks were tested for water absorption, acid resistance and compressive strength. A British Standard test (BS 1257:1945) gave 24 hour water absorption values, expressed in terms of weight, of 0·7–3·4 % for the four specimens that had been exposed to the atmosphere and 1·8–5·5 % for the four that had been immersed (and the resistance to acid tests (BS 784:1953) gave corresponding values of 99·1 and 98·9 %, based on one specimen in each case). Compressive strength tests (using two specimens that had been immersed and two exposed to the atmosphere) gave values ranging between 69·7 and 72·8 MN/m^2, apart from one of the immersed specimens which failed at 36·8 MN/m^2.

The report indicated a number of possible applications for which the slate-based fired bricks might have been considered; these included uses as highest grade engineering bricks, acid-resisting engineering bricks, chimney-lining bricks above fire level, abrasion-resistant paving (e.g. for food and fruit juice factories, slaughterhouses, dairies and bacon factories) as well as applications in water and canal works, sewers and underground tanks, and in the chemical industry.

Other opinions seemed to confirm the potential value of these bricks as far as their properties are concerned:[7] an industrial laboratory report (giving the results of tests for the acid resistance, water absorption, apparent porosity and specific gravity of a sample) indicated their suitability for acid-resistant construction work; one local authority would have used such bricks, had they been commercially available, for sewage and related works; and the condition of some bricks that had been built into the wall of a canal lock appeared to be unaffected after about 6 years' service.

However, the lack of plasticity of the slate mixes and the necessity for careful handling of the 'green' bricks could prove to be serious disadvantages in production. And the critical temperature control required to achieve vitrification without bloating could preclude the use of slate on a full-sized production scale;[8] furthermore, the market for acid-resisting brick is declining.

There have also been various publications which refer to the use of clay slate and the slate-like waste from coal production for brick making.[9–12] In the United Kingdom, bricks have been made from colliery spoil for many years[13] although production is now declining because surface clays are more satisfactory in a number of respects; it is interesting to note that a special process was developed because of the low plasticity associated with colliery spoil.

6.4. ALUMINA AND ALUMINIUM

In 1946, North[14] noted that an average slate, with an alumina content of 17–18%, contains the equivalent of 9% metallic aluminium. He pointed out that, although aluminium extraction could be carried out on a laboratory scale, it was not commercially feasible at that time. An earlier American patent[15] and a number of later publications[16–18] have referred to the extraction of alumina from slate and similar materials.

Recent worldwide events in the alumina and aluminium industries, described by Christie and Derry[19] and by Crockett,[20] have some bearing on the possibility of producing aluminium from slate. In Western countries, almost all aluminium is produced from bauxite; and generally, in approximate terms, 4 tonnes of bauxite are needed to provide the 2 tonnes of alumina from which 1 tonne of aluminium is made.[19] Bauxite production is tending to be concentrated in a decreasing number of countries[20] and, in 1974, major producing nations formed a cartel which has had some influence in the world market. Rising transport costs may encourage the production of alumina from bauxite in the producing countries, and perhaps the production of aluminium too.[19] In view of these factors, it is not surprising that the possibility of producing aluminium from indigenous sources has received careful attention in countries with no significant bauxite deposits.

Christie and Derry[19] have recently reviewed the possibilities of producing aluminium from sources within the United Kingdom. Their recommendations include the examination of processes which might possibly be used with colliery spoil, china clay micaceous waste and pulverized fuel ash as raw materials. They noted that these three materials appeared to be the most promising, in the United Kingdom, on the basis of alumina content, cost and availability; they commented on the relatively low alumina content of slate waste, and its chemical inertia, but noted its ready availability if a suitable process could be found.

6.5. GLASS PRODUCTS

The Mineral Industries Experiment Station Bulletin[2] described a brief investigation into the use of slate as a source of alumina in making amber glass; however, this work showed that the high content of carbon, iron oxide and pyrite in the slate was a disadvantage.

Melting tests, in which all the alumina requirement was provided by the slate, were carried out in a gas-fired furnace; and the melting and firing properties were found to be excellent but the colour was much too dark because of the impurities. Various oxidizing agents were used to improve the colour but it was noted that these would make the process too expensive. However, a good amber colour was achieved by using an electric furnace with a highly oxidizing atmosphere.

Other tests were carried out, in the gas-fired furnace, in which only part of the aluminium was supplied by the slate, the remainder being provided by the addition of aplite. The glass showed good melting behaviour but it was still too dark, although capable of being modified by oxidation.

The report concluded that, although the melting and firing behaviour of the slate was satisfactory, it would be necessary either to use the electric furnace method or to employ oxidizing agents to achieve the required colour; in the latter case, the price at the factory would have to be low enough ($2·76 per tonne at that time, in 1947) to allow the use of expensive oxidizing agents.

As an alternative approach, the possibility of benefication of the slate was suggested; for example, it might be possible to employ a flotation technique to remove some of the pyrite which seemed to be particularly deleterious.

There has also been some comparatively recent interest in slate as a glass making material.[21]

6.6. OTHER APPLICATIONS

Waste utilization can be achieved using processing techniques already employed by the slate industry. For instance, one quarry produced 102

× 102 mm tiles, prepacked for the do-it-yourself market, from material that would otherwise have been tipped as waste[22] and, at another quarry, walling stone has been produced by sawing waste block slate into pieces of appropriate sizes.[23] Also, slabs of slate waste have been used for river control and coastal defence work.[8]

REFERENCES

1. *Slag or Mineral Wool from Slate Waste*, Building Research Station Note No. B 163, Department of Scientific and Industrial Research (May 1956).
2. *Properties and New Uses of Pennsylvania Slate*, Mineral Industries Experiment Station Bulletin No. 47, Pennsylvania State College (1947).
3. 'Dinorwic—one of our great slate producers', *Mine and Quarry Engineering*, **15**(12) (Dec. 1949), 371–7.
4. H. Stamboliev and S. Kovacevski, *Tehnika* (*Belgrade*), **23**(3) (1968), 473–5. (*Chem. Abstr.*, **69** (1968), 4825v.)
5. G. O. Phillips (Executive Principal, The North E Wales Institute of Higher Education), private communication.
6. S. Kent, *The Production of Acid Resisting Engineering Bricks from Welsh Slate Waste*, Rural Industries Bureau, London (Oct. 1962).
7. F. S. Dexter (Brick Consultant, Council for Small Industries in Rural Areas), private communication.
8. W. Davies, *The Utilization of Slate Waste*, Report prepared for The Development Board for Rural Wales (Mar. 1978).
9. S. Pawlowski, *Proc. Conf. Silicate Ind., 6th, Budapest* (1961), (publ. 1963), 353–60. (*Chem. Abstr.*, **61** (1964), 5335g.)
10. P. Bayliss and F. C. Loughnan, *Clay Minerals Bull.*, **5** (1964), 353–62. (*Chem. Abstr.*, **63** (1965), 9694c.)
11. S. Bugrim, *Stroitel. Materialy*, **4**(2) (1958), 29–30. (*Chem. Abstr.*, **52** (1958), 15017c.)
12. K. Tyszka and A. Giezynski, *Chim. Ind.* (*Paris*), **93**(6) (1965), 614–21. (*Chem. Abstr.*, **65** (1966), 474h.)
13. W. Gutt, P. J. Nixon, M. A. Smith, W. H. Harrison and A. D. Russell, *A Survey of the Locations, Disposal and Prospective Uses of the Major Industrial By-products and Waste Materials*, Building Research Establishment Current Paper CP 19/74, Department of the Environment (Feb. 1974), 7.
14. F. J. North, *The Slates of Wales*, 3rd Edition, The National Museum of Wales (1946), 83–4.
15. US 2,248,826 (1941). (*Chem. Abstr.*, **35** (1941), 6745[6].)
16. K. Akerman, B. Żmudziński, Z. Orman and Z. Musienko, *Arch. Hutnictwa*, **1** (1956), 319–39. (*Chem. Abstr.*, **51** (1957), 12383b.)
17. Ger. Offen. 2,257,521 (1973). (*Chem. Abstr.*, **79** (1973), 68255y.)
18. H. E. Shafer, Jr, and J. A. Solomon, *Trans. Soc. Mining Eng. AIME*, **238**(1) (1967), 101–3. (*Chem. Abstr.*, **67** (1967), 4421u.)

19. P. Christie and R. Derry, *Aluminium from Indigenous UK Resources: A Review of Possibilities*, Warren Spring Laboratory Report No. LR 219 (ME) (1976).
20. R. N. Crockett, *Bauxite, Alumina and Aluminium*, Mineral Dossier No. 20, Mineral Resources Consultative Committee, HMSO (1978).
21. Czech. 156,280 (1974). (*Chem. Abstr.*, **83** (1975), 64854c.)
22. S. Herbert, 'Modernization of a Welsh slate quarry', *The Quarry Managers' Journal*, **56**(11) (Nov. 1972), 367–70.
23. 'The quarrying of Westmorland blue slate', *The Quarry Managers' Journal*, **48**(11) (Nov. 1964), 445–54.

7

Reclamation

Clearly the location and sometimes the composition of particular slate waste tips may preclude the possibility of using them as raw material for the manufacture of products. Nevertheless, on aesthetic grounds, it may be desirable to make them more visually acceptable, particularly where they are situated close to housing or in areas of scenic beauty. Under some circumstances, it may be possible to recontour them and there have been cases where it has been possible to put waste material into adjacent disused workings. Figure 7.1 illustrates one such scheme. Backfilling with waste is sometimes carried out during the course of commercial quarrying operations but large-scale recontouring can be an expensive operation that may be very difficult to justify simply on aesthetic grounds in the absence of other pressures.

Reclamation by vegetation is another possibility that is illustrated in Fig. 7.1. In discussing the rehabilitation of wastes from mineral extraction, Goodman[1] has listed various major groups of materials, including slate and shale wastes, and in each case has noted the most important factors which inhibit the growth of plant life upon them. In the case of slate waste, these inhibitory factors tend to be physical. For instance, Goodman has noted that problems associated with toxicity and spontaneous combustion do not arise; however, the material is deficient in nutrients as well as in micro-organisms and beneficial fauna such as earthworms. Also, the surface of slate waste tips can be steep and is generally uneven, usually without fine soil-forming material, and slippage tends to occur; adverse temperature regimes, wind turbulence and a variable supply of water are further, and sometimes related, factors which Goodman noted as tending to inhibit the establishment of plant growth.

The reclamation of slate waste tips by tree planting has been examined by Crompton,[2] Sheldon[3,4] and by Sheldon and Bradshaw;[5] it is clear that there is now some considerable practical understanding of the factors involved in the successful establishment of trees on slate waste.

(*a*) *Some associated abandoned housing and a tip at the north end of the quarry.*

(*b*) *and* (*c*) *The commencement of regrading work and partial filling at the north end of the quarry.*

Fig. 7.1. Recent reclamation work at the Coed Madog slate quarry in North Wales. (The quarry was abandoned in about 1910.) (Copyright photographs reproduced by courtesy of Dr C. G. Down.)

(*c*)

(*d*) *The south end of the quarry showing a slate tip, on the far side, which has been regraded and seeded some years earlier.*

Fig. 7.1—contd.

If the surface of a slate waste tip contains a high proportion of fine material, and if the moisture content is relatively high, then small plants can become established; various grasses grow under such conditions, particularly in flat areas and where the presence of animals, such as rabbits and sheep, ensures a supply of nutrients.[4] Mosses and ferns are also found. However, the surface of many tips is characterized by relatively large pieces of slate with considerable voids in between, although fine material often accumulates in the voids a short distance below the surface. (Such fine material may be formed by movement in the tip and by weathering processes, or it may be present in the form of dressing and mill waste.[4]) The inhospitable and generally moisture-deficient conditions at the surface of such tips inhibit the development of shallow-rooted plants such as grass, but deeper rooted plants can survive once they are established.

Trees establish themselves quite naturally on slate waste and grow there successfully although generally rather slowly; and there are examples of quarries where natural colonization has given them an acceptable appearance within about 50 years.[2] Natural colonization by trees tends to occur around the bottom of the tips, between them, and in drainage gullies, as well as on material containing a high proportion of fines.[3] Birch is one of the most successful species in moist conditions since it shows rapid growth and has modest nutrient requirements;[3,4] furthermore, when it is established it seeds well and flourishes under wet and dry conditions as well as providing leaf litter for the formation of humus—and, ultimately, it tends to become surrounded by dense undergrowth. Goat willow also flourishes under moist conditions. But some trees, such as oak and sycamore, occur towards the top of slate waste tips where surface conditions are more hostile;[4,5] even where the seeds of these species have fallen on to inhospitable areas of slate waste, the large food reserves which they contain and their rapidly developing roots have enabled them to quickly establish an adequate supply of water and nutrients for continued growth.

Once a sapling is established, then the production of leaf litter can provide humus for the colonization of other species;[3,4] and in this way, a self-sustaining plant community can develop. Furthermore, the accumulation of material in the voids tends to improve the local stability at the surface of the tip.[5] The main problem seems to lie in the establishment stage. Apart from water, which generally drains rapidly from the surface of slate tips, the developing tree needs nutrients. Slate waste cannot supply nitrogen, although this is provided by rainfall and by animals such as rabbits and sheep; but, unfortunately, the grazing habits of these animals, and sometimes human activity, tend to endanger the developing seedlings and suitable fencing is generally required.

Trees have been planted on slate waste, in paper pots containing earth, with varying degrees of success.[2] And, more recently, considerable success has been achieved by planting saplings in pockets of peat placed in waste tips.[3–5] Of the various media investigated, peat was found to be the most successful since it holds the water well and added nutrients are less liable to be leached out; furthermore, it is light to handle when working on the tips and disturbance, with consequent slippage of pieces of slate, can be minimized. It was found that the pocket of peat should be large enough to support the initial growth of the sapling but sufficiently small to encourage the roots to extend beyond it relatively quickly and thus enable the plant to become self-supporting.

The peat, to which appropriate quantities of slow release magnesium ammonium phosphate fertilizer and ground limestone are added,[5] can be placed by hand. In one area, local schoolchildren carried out a planting programme which resulted in the establishment of birch, lodgepole pine, mountain ash, red oak and sycamore;[4] such programmes could be carried out at little more cost than that of the saplings and the compost. However, there are some advantages to be gained by pumping the peat as a slurry, containing wood pulp for bulk, from a tanker;[5] this technique can reduce the amount of disturbance to the tip caused by walking on it. Furthermore, the slurry penetrates down to the fine waste material, and planting can be carried out shortly after pumping (without waiting for rain to moisten the peat as in the case of hand placing). In 1976, the cost of planting one tree (including the sapling and other materials as well as plant hire and excavation of the pocket) was estimated to be about £1;[5] fencing and further maintenance would cost extra.

To illustrate the aesthetic benefit that vegetation can produce, Fig. 7.2 (a)–(e) shows some examples in Wales.[6] Figure 7.2(a) illustrates the natural colonization of willow on a slate tip in North Wales. Figure 7.2(b) shows the natural colonization by birch, rhododendron and heather on a tip slope in mid-Wales that consists of a relatively soft slate which is vegetated comparatively readily because it weathers more easily to a soil-like material. Figure 7.2(c) illustrates birch colonization on hard slate waste tips at Bethesda in North Wales; because of the limitation of water, growth begins in the valleys where conditions are most suitable and then spreads slowly as leaf litter improves the conditions for seedling establishment. Figure 7.2(d) shows alder transplants in an experiment to test tree establishment techniques on slate waste in Bethesda. Alder has the great advantage of being able to fix its own nitrogen, thus overcoming the problem of nutrient deficiency;[4,5] and planting alder may be a means of increasing the proportion of nitrogen in the waste for the benefit of other species. Figure 7.2(e) shows a lodgepole pine, established in a peat pocket, in an experiment to test planting and establishment techniques on a slate waste slope in Bethesda. Although lodgepole pine grows well, its leaves decay slowly and therefore it does not create soil successfully;[4,5] it would seem to be of limited value unless commercial exploitation were possible. (An investigation has been carried out into the composition of soil formed from weathered slate, among other materials, and litter from blue pine seedlings.[7])

The policy for maintaining saplings on slate waste still needs to be established,[5] although these preliminary investigations suggest a relatively cheap approach to the aesthetic problems associated with slate waste tips that cannot be used otherwise. (A related possibility that has been suggested by Johnson, McNeilly and Putwain[8] is the use of slate waste as a surface amendment applied to spoil heaps which contain toxic elements, such as those from mining non-ferrous metal ores. It appears that inert amendments are suitable for aesthetic revegetation; with more 'fertile' amendments, the roots rapidly grow through into the toxic substrate. Further investigations using slate waste have been planned.[9])

Apart from the waste itself, disused workings and associated buildings can sometimes be put to good use. An abandoned slate quarry, at Wind

(a)

Fig. 7.2. The growth of vegetation on slate waste (by courtesy of Dr J. C. Sheldon).

(b)

(c)

(d)

(e)

Fig. 7.2—contd.

Gap in Pennsylvania, was adapted for use as an oil store with a capacity of one million barrels;[10] the cost of the store, per barrel capacity, was said to be a quarter of that of a conventional tank storage scheme and the area which it covered was considerably less. The surface of the oil, which was covered by a floating raft of steel pontoons, was maintained at a constant level by pumping water into or out of the bottom of the quarry, underneath the oil, as required; and it was anticipated that the oil would not seep out of this particular quarry since it was kept below the level of the surrounding water table. In another scheme, in North Wales, it was proposed to flood disused slate workings for fishing and sailing and to landscape waste tips and build hotels and houses.[11] Underground slate mines have been used for storage purposes. These are sometimes situated in remote locations which can make them particularly useful as explosives and high-security stores; one such mine in the United Kingdom was used for the storage of art treasures and the Crown Jewels during the Second World War.[12]

The presence of slate workings in a tourist area can be turned to good advantage, and two enterprises in North Wales enable the visitor to understand something of the technological and social history of the industry for which that area is particularly well known. The Dinorwic Quarry closed in 1969 and the local County Council bought the quarry workshops (built in 1870) which are now maintained by the Department of the Environment and operated as a museum by the National Museum of Wales.[13] The museum contains a large collection of machinery and tools associated with many aspects of the manufacture and maintenance of equipment that was used in the North Wales slate industry; there are also exhibits which illustrate slate quarrying and finishing, and others which reflect the social history of the industry. In 1972, the Llechwedd Slate Caverns at Blaenau Ffestiniog were opened to the public.[14] The visitor can make a railway tour through underground slate workings which contain tableaux demonstrating methods and equipment that were used in Victorian times; there is also a museum where early machinery and other exhibits can be seen. Shortly after opening, this enterprise won two major awards for tourism and is currently attracting about a quarter of a million visitors each year.[15]

REFERENCES

1. G. T. Goodman, 'Ecology and the problems of rehabilitating wastes from mineral extraction', *Proceedings of the Royal Society of London*, *A*, **339**(1618) (Aug. 1974), 373–87.

2. C. T. Crompton, 'The treatment of waste slate heaps', *Town Planning Review*, **37** (Jan. 1967), 291–304.
3. J. C. Sheldon, 'The growth of amenity trees in difficult conditions', *Proceedings of the Conference on Tree Growth in the Landscape* (Editors: S. E. Wright, G. P. Buckley and T. Wright), Wye College, Kent (1974), 69–76.
4. J. C. Sheldon, 'The reclamation of slate waste', *Nature in Wales*, **14**(3) (Mar. 1975), 160–8.
5. J. C. Sheldon and A. D. Bradshaw, 'Techniques No. 23: The reclamation of slate waste tips by tree planting', *Landscape Design*, No. 113 (Feb. 1976), 31–3.
6. The photographs and their description were provided by Dr J. C. Sheldon (Regional Ecologist, Department of Physical Planning, Lothian Regional Council).
7. M. N. Malik and M. J. Rehman, 'Some chemical characteristics of soil profiles formed from the weathering of limestone, shale and slate, and their effect on the nutrient composition of humus, litter and needles from seedlings of blue pine', *The Pakistan Journal of Forestry*, **20** (Apr. 1970), 189–99.
8. M. S. Johnson, T. McNeilly and P. D. Putwain, 'Revegetation of metalliferous mine spoil contaminated by lead and zinc', *Environmental Pollution*, **12**(4) (Apr. 1977), 261–77.
9. M. S. Johnson (University of Liverpool), private communication.
10. F. D. Hoyt, *What's New in the Pennsylvania Slate Industry*, Mineral Industries Experiment Station Bulletin No. 66, The Pennsylvania State University (1956), 49.
11. R. N. Crockett, *Slate*, Mineral Dossier No. 12, Mineral Resources Consultative Committee, HMSO, London (1975), 9.
12. C. G. Down and J. Stocks, *Environmental Impact of Mining*, Applied Science Publishers Ltd, London (1977), 237.
13. D. Morgan Rees, *North Wales Quarrying Museum*, Department of the Environment Official Handbook, HMSO, London, 2nd Impression (1976).
14. I. Wynne Jones, *Llechwedd Slate Caverns*, Quarry Tours Ltd, Blaenau Ffestiniog (1976).
15. I. E. Rutherford (Managing Director, Llechwedd Slate Caverns), private communication

8

The Broader Issues

8.1. INTRODUCTION

Earlier chapters have examined the technology of slate waste utilization, and the previous chapter has discussed reclamation as a solution to aesthetic problems where utilization is not feasible. The present chapter discusses broader aspects from several points of view. First, the benefits of utilization are considered, which include not only those to the slate industry and to potential users of slate waste products, but to the community at large. The factors which tend to inhibit more widespread utilization are then examined, together with considerations which may tend to offset them.

8.2. THE BENEFITS OF SLATE WASTE UTILIZATION

For a long time, the slate industry has been aware that the profitable use of stockpiled waste could help to offset the effects of its overall decline. The current cost of waste disposal, including the cost of its extraction, may be well in excess of £1 per tonne in some cases. As Davies[1] has pointed out, this contributes towards the price of the finished products and, where the wastage rate is high, the oncost is very significant. If waste disposal could be avoided, then the products would be cheaper and more competitive; and if this could be achieved by profitable utilization, then the benefit would be further increased.

The general advantage to potential consumers of slate waste products is no less apparent. The industrial community is consuming increasing quantities of raw materials at the same time as seeking to conserve its dwindling resources. This paradox can be partly resolved by the more efficient recycling of industrial by-products and waste.

Although the aesthetic benefits of removing slate waste tips are difficult to quantify in financial terms, they are still obvious. Furthermore, the processing of waste in slate-producing areas would contribute to the local economy; and, in some regions, it would help to solve unemployment problems.

Many slate waste products could contribute towards energy saving. For instance, the very high consumption of energy involved in manufacturing plastics (about 160 MJ/kg)[2] is diluted by using fillers in products made from them. To take another example, the energy consumed in producing lightweight aggregate by the rotary kiln process is relatively low, and may vary between 2·3 and 5·8 MJ/kg depending on the thermal efficiency of the kiln,[3] compared with 8 MJ/kg for cement manufacture and 48 MJ/kg for the production of finished steel.[2] Although due caution must be exercised in comparing the energy cost of engineering materials on a weight basis alone,[2] the comparison of alternative approaches to design in building and civil engineering construction (in terms of energy consumption in the methods and materials involved) is an increasingly significant consideration. The greater use of lightweight aggregates by the construction industry could lead to a reduction in the quantity of other materials used, as well as to lower transportation and handling costs. Furthermore, the improved thermal performance of lightweight aggregate concrete leads to direct energy savings during the lifetime of a building constructed from it. Similar arguments apply to the more widespread use of autoclaved aerated concrete, and slate wool is another product which has potential for saving energy.

The energy and human effort expended in extracting the slate waste has already been paid for, as a component in the cost of the products, but the material still remains. By making use of it now, other resources are conserved as well as the energy which would be consumed in extracting them; further dereliction (or the cost of preventing it) is avoided at the same time as reducing that arising from the slate waste.

Despite these rather obvious incentives, slate waste is still produced and there is very little utilization of existing stockpiles. This raises the question of what inhibitory factors are operating and what steps must be taken to overcome them.

8.3. FACTORS WHICH TEND TO INHIBIT SLATE WASTE UTILIZATION

It would seem to be a fairly straightforward exercise to evaluate slate as a resource in relation to its various potential applications. For instance, it is not difficult to estimate the value of the aluminium which a particular slate contains and the cost of extracting it (although, in this case, such calculations are unlikely to be very encouraging at the present time). The main problem seems to lie in viewing the plant and production costs against the complex background of environmental, social, geographical and general economic factors involved.

A specific problem that is often quoted is the distance of many slate waste sources from potential markets. The manufacture of fired bricks and autoclaved slate/lime bricks could not be undertaken successfully unless the transportation costs allow a competitive price. An enterprise of this kind would be unlikely to succeed in North Wales, for example, where the cost of reaching a market of adequate size would be too high. Changes in local circumstances may sometimes help to overcome the transportation problem; for instance, the oil-related developments in Scotland could possibly offer prospects for the utilization of tips in that area[4] although this would require careful evaluation.

The key to the proper analysis of the location effect lies in carefully balancing the variables involved in determining the maximum economic haulage distance. A recent report[5] on the use of waste materials as potential replacements for highway aggregates in the United States has suggested that, for this application, the limits are about 40 miles for road transportation and 100 miles and 300 miles for rail and barge respectively. Obviously, a specific proposal may depend on convenient access to a railway line or to a waterway, but due account must also be taken of factors such as onloading and offloading costs as well as the necessity, or otherwise, of having to make the journey in several stages by different means of transport. Pipeline transport is another possibility that might be considered under some circumstances; for instance, a slurry pipeline could be used to transport powder from a remote quarry to an autoclaved aerated concrete plant situated close to a railway line. The form in which the material is transported is another consideration. Expanded slate might be produced where the waste is situated or, alternatively, the raw material might be transported in suitably crushed form to a processing plant located close to the market.

The maximum economic transport distance also depends on the nature of the product and the processing costs which have gone into it. In the case of crushed slate for use as road fill or low grade aggregate in concrete, this may be considerably less than 40 miles, depending on local circumstances. It may be argued that this suggests that there is little prospect for making widespread use of tips in this way. However, it should not be forgotten that the total area encompassed, for the sake of argument, by 30-mile radii drawn around every substantial tip in existence must be quite considerable; and the availability of portable processing plant[5] may affect the extent to which small, local tips can be used.

Powder and granules are more costly and can be transported economically over larger distances; for instance, these products are exported from North Wales to the European Continent.

The potential sales of expanded slate in a given area would, of course, be highly dependent on the availability of alternative materials which are themselves often produced from other wastes; again, the availability of portable processing plant might enable advantage to be taken of localized shortages. It has been suggested that the potential market for expanded slate is confined to within a radius of about 150 miles of the plant, but there are instances where the economic transport distance has proved to be much further. For example, a five-storey building near London has been built from precast structural components made with expanded slate from North Wales. Furthermore, a hotel in Gibraltar[6] has been constructed using precast and prestressed expanded slate concrete components that were manufactured 1400 miles away, in the United Kingdom, and shipped to the site.

This raises the interesting possibility of not only producing expanded slate but also using it to manufacture constructional units in areas where the waste is located; by raising the proportion of the manufacturing component in the total cost of the product, the maximum economic transport distance could perhaps be extended considerably. The same argument might also be applicable to the production of precast units made from autoclaved aerated concrete. This approach has the additional benefit of improving the economy in areas which have suffered as a result of the decline of the slate industry.

Apart from transportation, there are other considerations. It has been suggested that there is a stigma attached to the word 'waste';[5] furthermore, engineers are understandably happier working with established materials of known performance characteristics than with comparatively unknown materials which may give rise to unanticipated problems. It has also been said that specifications and codes of practice tend to be restrictive when the use of waste materials is envisaged. On the other hand, overenthusiasm of research workers and others anxious to promote waste utilization has sometimes led to exaggerated or misleading claims and overstatement of the benefits to be obtained.

There are also difficulties in quantifying the more general benefits, particularly to the areas in which slate waste is located. It is not too difficult to estimate the effect of reduced unemployment, or even the overall economic advantage to the local community, but it is not easy to quantify the aesthetic benefits. Any proposal to set up facilities to process and transport slate waste products is likely to encounter some resistance on environmental grounds, particularly when reclamation seems to be a possible alternative. In view of this, and the health hazard associated with slate dust, it is essential to ensure that proper attention is paid to environmental and safety aspects.

The shortage of lightweight aggregates in the United States after the Second World War focused some attention on slate waste. The current resource problems, which are now more fundamental and widespread, are having the same effect. However, the extent to which utilization can contribute towards the solution of these problems depends not only on individual efforts but on co-operation; narrow viewpoints and lack of foresight seem to be major obstacles.

8.4. THE FUTURE

In 1960, a Building Research Station report[7] concluded that the limitations to slate waste utilization stemmed from economic and geographical factors rather than inadequate research effort or a lack of initiative in contemplating new uses. It is certainly true that many of the applications, both actual and potential, that have been discussed in this book had been recognized at that time. It is also true that there is no evidence to suggest that the slate industry as a whole has failed to recognize new opportunities although, due to its decline, individual companies have not always been able to invest in them. Without further investment, it is difficult to see what steps could be taken in the future. Sales in traditional markets have suffered with the introduction of alternative materials and there seems to be little or no prospect of reversing the overall trend, although a small number of enterprises can still profit from the remaining demand. There may, however, be some advantage in reappraising the market for slate as a filler and extender; a few companies who have already invested in the necessary plant find profitable outlets in a variety of industries. Compared with many fillers, slate powder is relatively cheap and, although its colour may be a disadvantage for some applications, cost and technical performance are often the primary considerations. Thorough market research, with appropriate supporting technical investigations, might identify applications in which it could replace more expensive fillers, such as talc and mica, for example.

The commercial development of new products and applications could perhaps, in some cases, be made easier by the active co-operation of other industries. For instance, in the United Kingdom, mineral wool production is carried out by a relatively small number of companies with the advantage of specialized knowledge of the technology and markets. Commercial developments should also take into account all the possibilities; expanded slate has been used as a concrete aggregate and as a filter medium, but perhaps it may have potential for other applications.

Returning to the Building Research Establishment report, economic and geographical as well as other constraints still appear to operate; but rapidly rising prices, resource problems, and sometimes environmental and even political pressures have reached the point where the validity of these constraints should perhaps be reinvestigated. The interactions between the various pressures and constraints is extremely complex and the development of successful policies for more effective utilization is likely to depend on the collaboration of industry and government, at least in the United Kingdom. There is also a more fundamental problem which seems to call for government involvement; the long-term and sometimes nebulous benefits of utilization are not necessarily reflected in the direct and relatively short-term profitability normally required of a commercial operation. A specific proposal may not appear to be commercially attractive when viewed in terms of the usual financial criteria, although there may be considerable overall benefit to the community as a whole.

This implies that, in the United Kingdom and probably some other countries, there is a case for investigating ways in which increased public interest could stimulate the commercial sector of industry. It is not suggested that public money must be spent to provide plant or to develop transport facilities although, in some cases, there may be justification for this kind of support where the benefits are obvious and readily quantifiable. A generally more satisfactory approach is to create a more responsive environment. To take an example, the dual tendering system, used for new road contracts in the United Kingdom, involves the assessment of using wastes as alternatives to other materials for bulk fill.[8] Efforts will certainly be made to promote the more widespread use of constructional materials which would enable energy resources to be conserved; this will provide impetus to the commercial development of some slate waste products. But, in order to develop new uses or to extend existing ones, it is necessary to overcome the effects of some of the other inhibitory factors discussed in the previous section. Engineers will only use materials if they have sufficient confidence in them; and committees which are responsible for drafting standards will only revise them if there is sufficient evidence to show that this can be done safely. Apart from this, in contemplating any application for slate waste from a particular source, it is essential to thoroughly investigate all relevant aspects of its use and performance. The necessary research and development work can be expensive; but publicly financed academic institutions and research establishments, with their specialized knowledge and facilities, could provide a great deal of technical support.

REFERENCES

1. W. Davies, *The Utilization of Slate Waste*, Report prepared for The Development Board for Rural Wales (Mar. 1978).
2. C. Hall, 'Chemistry and construction', *Chemistry in Britain*, **12**(10) (Oct. 1976), 317–20.
3. N. W. Biege and S. M. Cohen, 'Cut fuel costs in the lightweight aggregate industry', *Ceramic Bulletin*, **54**(6) (1975), 569–70.
4. E. M. Lawson and P. J. Nixon, *A Survey of the Locations, Disposal and Prospective Uses of the Major Industrial By-products and Waste Materials in Scotland*, Building Research Establishment Current Paper CP 50/78, Department of the Environment (Jun. 1978).
5. R. H. Miller and R. J. Collins, *Waste Materials as Potential Replacements for Highway Aggregates*, National Cooperative Highway Research Program Report 166, Transportation Research Board, Washington, D.C. (1976).
6. B. H. Spratt, *The Structural Use of Lightweight Aggregate Concrete*, Cement and Concrete Association Publication 45.023 (Dec. 1974).
7. *The Utilization of Slate and Slate Waste*, Building Research Station Note No. A 84, Department of Scientific and Industrial Research (Jun. 1960).
8. C. G. Down and J. Stocks, *Environmental Impact of Mining*, Applied Science Publishers Ltd, London (1977), 308.

Index